4th International Conference on Advances in Materials, Machinery, Electronics (AMME 2021)

MATEC Web of Conferences Volume 353 (2021)

Wuhan, China
23 - 24 October 2021

Editors:

Lin Liu
Fang Liu

ISBN: 978-1-7138-4183-8

Printed from e-media with permission by:

Curran Associates, Inc.
57 Morehouse Lane
Red Hook, NY 12571

Some format issues inherent in the e-media version may also appear in this print version.

For additional information, please contact EDP Sciences – Web of Conferences at the address below.

EDP Sciences – Web of Conferences
17, Avenue du Hoggar
Parc d'Activité de Courtabœuf
BP 112
F-91944 Les Ulis Cedex A
France

Phone: +33 (0) 1 69 18 75 75
Fax: +33 (0) 1 69 28 84 91

contact-edps@webofconferences.org

Additional copies of this publication are available from:

Curran Associates, Inc.
57 Morehouse Lane
Red Hook, NY 12571 USA
Phone: 845-758-0400
Fax: 845-758-2633
Email: curran@proceedings.com
Web: www.proceedings.com

TABLE OF CONTENTS

 https://doi.org/10.1051/matecconf/202135300001

Preface: 4th International Conference on Advances in Materials, Machinery, Electronics (AMME 2021)

This conference proceedings includes original and peer-reviewed research papers from the 4th International Conference on Advances in Materials, Machinery, Electronics (AMME 2021), and held on October 23-24, 2021 in Wuhan, China.

The aims of the conference are to provide an interactive platform for the academicians, R & D organizations, industries, innovators, entrepreneurs, government agencies and policy-makers etc., from both China and abroad to exchange ideas on recent researches and advances in the domain of materials, machinery, electronics with its prospective application in the various interdisciplinary domains of engineering. This two-day conference is being organized in technical association with Huazhong University of Science and Technology and will include invited keynote talks and oral paper presentations from both academia and industry to ignite our young minds, to initiate them in the field of meaningful research and thereby enrich their existing knowledge.

All full paper submissions to the AMME 2021 must be written in English and will be sent to at least two reviewers and evaluated based on originality, technical or research content, correctness, relevance to conference, contributions, and readability. The full paper submissions will be chosen based on technical merit, interest, applicability, and how well they fit a coherent and balanced technical program. AMME 2021 has received 61 manuscripts. And 26 submissions have been accepted by our reviewers. The Committees of AMME 2021 invest great efforts in reviewing the papers submitted to the conference and organizing the sessions to enable the participants to gain maximum benefit.

Hopefully, all participants and other interested readers benefit scientifically from the proceedings and find it stimulating in the process.

With our warmest regards,

Lin Liu, Fang Liu

Wuhan, China

Statement of Peer review

In submitting conference proceedings to *Web of Conferences*, the editors of the proceedings certify to the Publisher that

1. They adhere to its **Policy on Publishing Integrity** in order to safeguard good scientific practice in publishing.
2. All articles have been subjected to peer review administered by the proceedings editors.
3. Reviews have been conducted by expert referees, who have been requested to provide unbiased and constructive comments aimed, whenever possible, at improving the work.
4. Proceedings editors have taken all reasonable steps to ensure the quality of the materials they publish and their decision to accept or reject a paper for publication has been based only on the merits of the work and the relevance to the journal.

Title, date and place of the conference

4th International Conference on Advances in Materials, Machinery, Electronics (AMME 2021)
October 23-24, 2021
Wuhan City, China

Proceedings editor(s):

Lin Liu
Fang Liu

Date and editor's signature October 28, 2021

17, avenue du Hoggar - PA de Courtabœuf – BP 112 - 91944 Les Ulis Cedex A (France)
Tél. : 33 (0)1 69 18 75 75 - Fax : 33(0)1 69 07 45 17 – www.edpsciences.org

Research on Preheating Temperature Field Test of Natural Gas Pipeline in Service Welding under Extreme Condition

Yan Xu [1, *], *Yinglai Liu* [1], *Zhenjun Feng* [1], *Xianghui Nie* [1], *Chao Zhang* [1], *Jasmine* [2]

[1] Tubular Goods Research Institute of CNPC, Xi'an, China
[2] Companhia Brasileira de Metalurgia e Minerao, Brazil

Abstract. When the natural gas pipeline is welding in service, the fast flowing medium with pressure in the pipe will take away a lot of heat, and the preheating temperature is not easy to be guaranteed, so it is easy to appear hydrogen-induced crack. In this paper, the in-service welding preheating temperature field of natural gas pipeline under the limit condition of unreduced volume was studied, and the pre-welding preheating test was carried out by using the medium frequency heating method. It is found that the temperature below the heating belt increases gradually with the increase of the intermediate frequency heating power, and the fitting shows a quadratic polynomial gradient. There are differences in preheating temperatures on the same circumference. The highest temperature mostly appears in the direction of 3 point of the pipeline, while the lowest temperature mostly appears in the direction of 0 point, which is related to the tightness of the heating belt, sunshine, blowing and other factors. In addition, the preheating temperature field of the pipeline in service is related to the gas flow in the pipeline. At the same heating power, the downstream temperature of the heating belt is higher than the upstream temperature at the same location, and the closer to the heating belt, the higher the temperature is. When the gas flow rate reaches 9.37m/s and the heating power is 160kW, the average measured temperature at 50mm upstream and downstream of the heating belt of Φ1016 pipeline is 107℃, and the average measured temperature at 50mm upstream of the heating belt is 71℃. When the gas flow rate reaches 8.91m/s and the heating power is 200kW, the average measured temperature at the downstream 50mm of the heating belt of Φ1219 pipeline is 72℃, the average measured temperature at the upstream 50mm of the heating belt is 52℃ and the average measured temperature at the upstream 30mm of the heating belt is 71℃..

1 Introduction

As an important means of in-service welding repair, B sleeve repair technology can not only ensure continuous pipeline transportation, but also reduce the degree of environmental damage, shorten the repair cycle and reduce the repair cost [1-2]. The repair technology is applicable to a wide range of defects, including pipeline corrosion, cracks, mechanical damage, weld defects, etc., and can also repair leakage defects, with good repair effect and high reliability, which belongs to permanent repair[3-4].When the B-type sleeve is welded, the pipeline is in an uninterrupted state, and there is a rapid flow of pressurized gas or liquid medium in the pipeline, which is difficult to ensure the preheating temperature and interlayer temperature. At the same time, the cooling rate of weld is too fast, which will promote the formation of hydrogen-induced crack-sensitive harden structure in welded joints, so hydrogen-induced cracks are easy to occur and the bearing capacity of the joints is reduced. [5-6].Preheating before welding can slow down the cooling rate after welding, and promote the diffusion of hydrogen, water and other pollutants. It can not only avoid the generation of hydrogen-induced cracks, but also reduce the hardening degree of weld and heat affected zone, and improve the crack resistance of welded joints [7-11]. At present, the commonly used preheating methods before welding are flame heating and intermediate frequency heating, but the preheating temperature often cannot meet the welding requirements in the construction. The pipeline to be repaired needs to reduce the flow rate of the conveying medium. Therefore, it is necessary to study the relationship between the preheating temperature of the in-service welding of the natural gas pipeline and the heating power and the flow rate of the medium, especially the distribution of the preheating temperature field of the in-service welding of the natural gas pipeline under the extreme working conditions.

2 Testing method

Taking the West-East Gas Pipeline as the research object, several gas pipelines with Φ1016mm (wall thickness 14.6mm, pipe flow rate 4.60~9.82m/s) and Φ1219mm

*Corresponding author e-mail: xuyan008@cnpc.com.cn

(wall thickness 15.3mm or 18.4mm, pipe flow rate 7.51~8.91m/s) in service were excavated, and the anticorrosive layer was stripped. The pipe body was heated by medium frequency heating method, and the temperature distribution below and upstream and downstream of the medium frequency heating zone was recorded by thermocouple real-time test (Figure 1, the temperature test point number below the heating zone was P1~P16, and the temperature test point number above the heating zone was B1~B16). The temperature tracking was carried out at the test point 50mm away from the edge of the heating zone (the farthest welding channel position during B sleeve welding). The heating test was carried out by using several intermediate frequency heating equipment with rated power commonly used in welding. Different initial power was set according to different equipment. After holding for 10 min, it was increased by 20kW in turn. The heating was stopped when the preheating temperature at the farthest weld position reached the welding requirement (60°C) or the equipment power reached the limit.

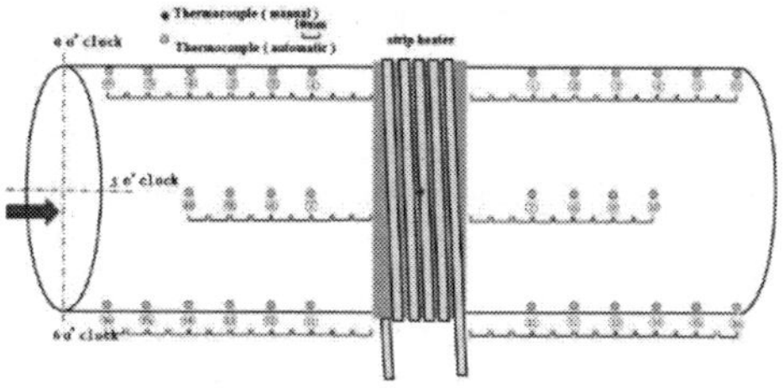

Figure 1. Layout of temperature test points

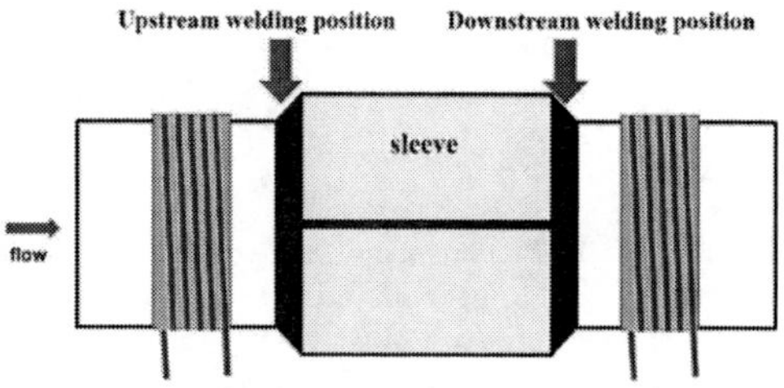

Figure 2. B sleeve welding schematic

3 Test results and analysis

3.1 Test temperature below tropical zone

The test temperatures of Φ1016mm pipeline and Φ1219mm pipeline with P1~P16 below the tropical zone under different heating power are summarized respectively (Figure 3). It can be seen that the temperature below the tropical zone of the pipeline and the heating power of the intermediate frequency equipment show a quadratic polynomial gradual change. With the increase of the heating power of the intermediate frequency equipment, the temperature below the tropical zone increases gradually.

$y=-4.04847+1.02761x+0.0046x^2$

$y=-1.40363+1.12687x+0.00321x^2$

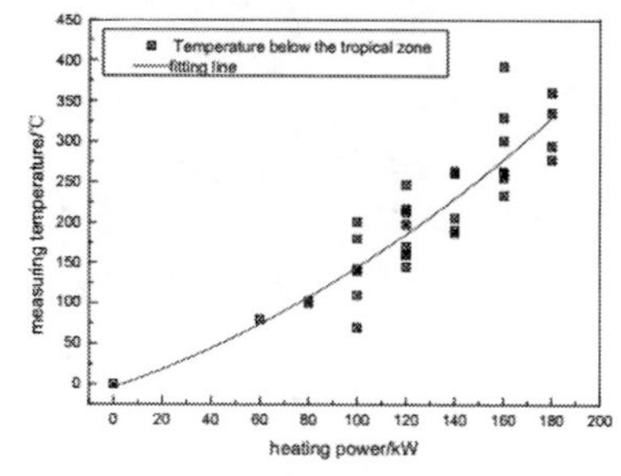

(a) Φ1016mm pipeline

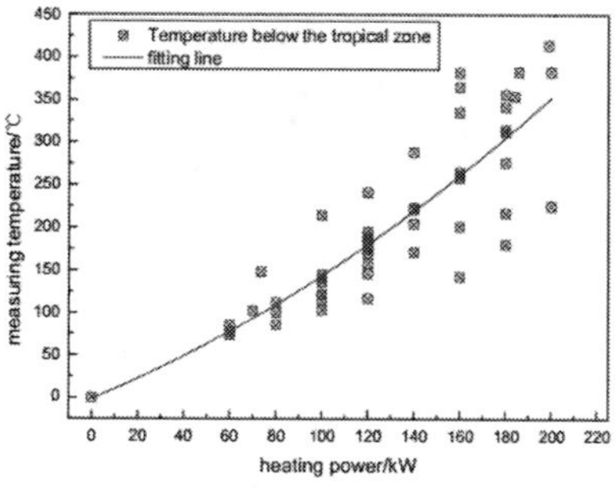

(b) Φ1219mm pipeline

Figure 3. The relation curve between the test temperature and the intermediate frequency heating power under the heating zone of different pipe diameters

3.2 Upper and Lower Tropical Temperatures

The heating temperature measurement of Φ1016mm pipeline under the condition of pressure 8MPa, flow rate 2850 × 104 m3/d and flow rate 4.60m/s was analyzed, and the temperature distribution and temperature difference curve of the upstream and downstream of the heating zone under different heating power were obtained (Figure 4). It can be seen that when the heating power is 180kW, the temperature below the tropical reaches 352°C, the upstream and downstream are closer to the edge of the tropical, the higher the temperature, the downstream temperature is significantly higher than the upstream temperature with equal spacing, and the maximum temperature difference is close to 40°C.

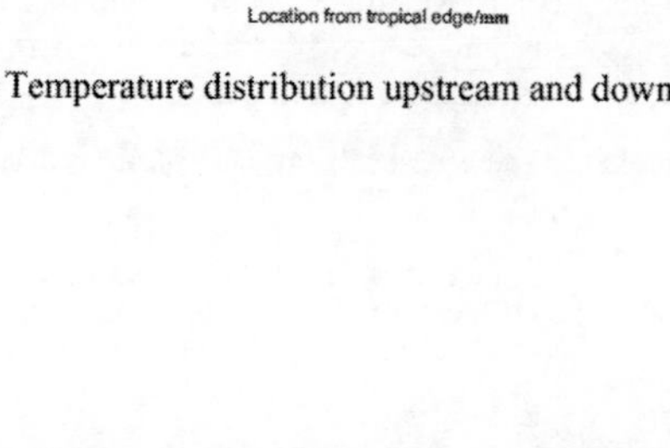

(a) Temperature distribution upstream and downstream

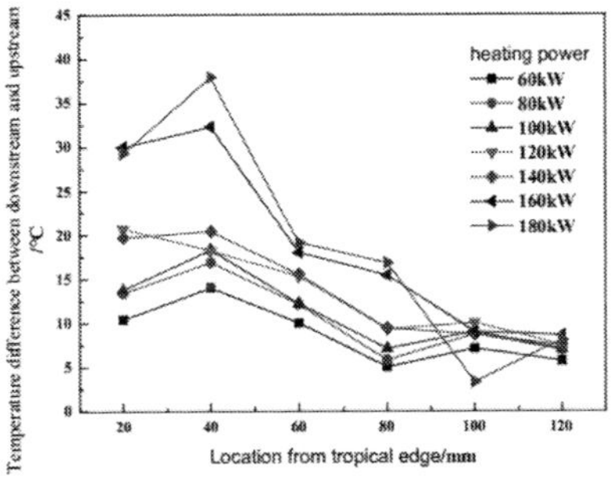

(b) Temperature difference upstream and downstream

Figure 4. Temperature distribution and temperature difference curves of the heating zone upstream and downstream under different heating powers of Φ1016mm pipeline

3.3 Different quadrant temperature tests

Under the ideal heating condition, the temperature distribution on the same circumference of the pipeline should be consistent, but in the actual measurement, it is found that there are differences in the data of 0, 3 and 6 points on the same circumference of the tropical zone, so as to ensure that other basic conditions remain unchanged. The temperature of Φ1016mm pipeline and Φ1219mm pipeline in different quadrants are tested respectively, and the temperature test results of different quadrants in the heating test of the pipeline under the limit working condition of the West-East Gas Pipeline are obtained (Figure 5). It can be seen that the maximum temperature mostly appears in the direction of 3 points, and the minimum temperature mostly appears in the direction of 0 points, which is related to the factors such as tropical winding tightness, sunshine and blowing. It is suggested that the heating effect should be improved as much as possible, and the heating coil should be tightly wrapped, especially at 6 o ' clock. Building windbreaks when necessary.

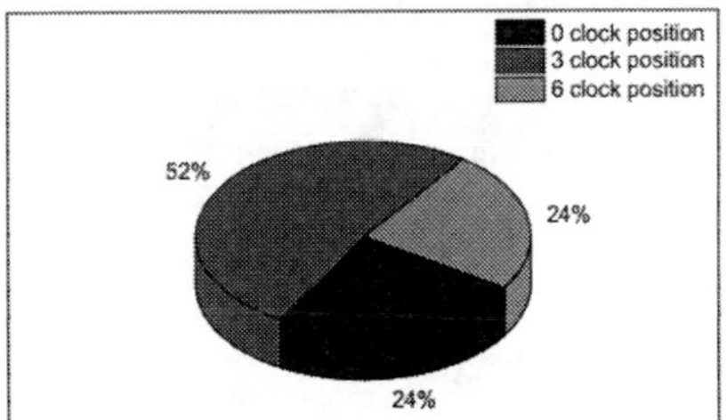

(a) The position of the clock at the highest temperature

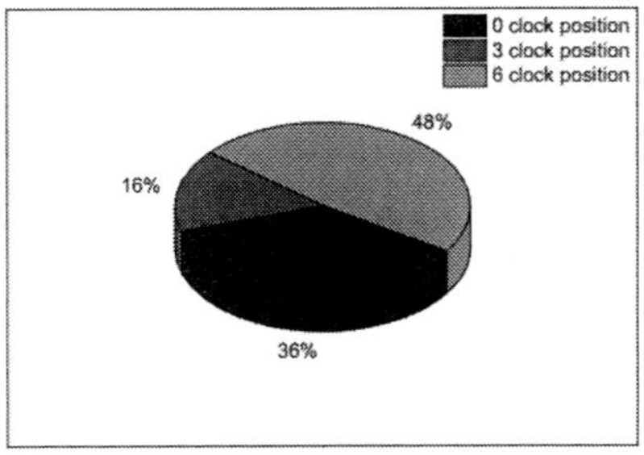

(b) The position of the clock at the lowest temperature

Figure 5. Test results of different quadrant temperatures for heating test of West-East Gas Pipeline

3.4 Analysis of preheating temperature of pipeline in-service welding under extreme conditions

According to the field feedback, the maximum gas flow rate of Φ1016 pipeline test section is 7-8m/s under normal operation, and the maximum gas flow rate of Φ1219 pipeline test section is 9-10m/s under normal operation. In this test, the maximum gas flow rate of Φ1016 pipeline reaches 9.37m/s, and the maximum gas flow rate of Φ1219 pipeline reaches 8.91m/s, which is close to or reaches the limit flow rate of the pipeline. Table 1 and table 2 are the test results of in-service welding preheating temperature of Φ1016 pipeline and Φ1219 pipeline under extreme conditions. The preheating temperature field of the in-service pipeline is related to the gas flow in the pipeline. Under the same heating power, the downstream temperature of the tropical zone is higher than the upstream temperature at the same position, and the closer to the tropical zone, the higher the temperature. When the gas flow rate reaches 9.37m/s, under the heating power of 160 kilowatts, the measured average temperature at 50mm downstream of the Φ1016 pipeline plus the tropical zone is 107°C, and the measured average temperature at 50mm upstream of the tropical zone is 71°C, which basically meets the preheating temperature (60°C) required by the current site. When the gas flow rate reaches 8.91m/s, under the heating power of 200kW, the measured average temperature at 50mm downstream of the Φ1219 pipeline heating tropical zone is 72°C, and the measured average temperature at 50mm upstream of the heating tropical zone is 52°C. Among them, the preheating temperature at 50mm upstream does not meet the requirements of welding preheating, while the measured average temperature at 30mm upstream of the heating tropical zone is 71°C. Therefore, it is recommended that the B-type sleeve should be installed in the Φ1219 pipeline under this condition, and the methods simultaneous heating of upstream and downstream, moving heating tropical zone and auxiliary heating should be used to meet the requirements of welding preheating.

https://doi.org/10.1051/matecconf/202135301001

Table 1. In-service welding preheating temperature test results of Φ1016mm pipeline and Φ1219mm pipeline under different equipment output conditions under limit conditions

pipe diameter / mm	flow velocity / ($m \cdot s^{-1}$)	heating power /kW	Measured average temperature /°C			
			Under the tropical zone	Upper tropics 50mm	Upper tropics 30mm	Lower tropical zone 50mm
1016	9.37	80	103	49	82	66
		100	172	56	99	80
		120	212	59	106	88
		140	310	69	135	103
		160	330	71	145	107
1219	8.91	100	145	50	48	50
		120	183	53	52	54
		140	221	49	56	58
		160	265	52	60	66
		180	312	53	66	69
		200	383	57	71	72

4 Conclusion

(1) The temperature below the heating band gradually increases with the increase of the heating power of the intermediate frequency equipment, and the fitting shows a quadratic polynomial gradual change.
(2) The preheating temperature on the same circle is different, which is related to the tropical winding tightness, sunshine, blowing and other factors. Among them, the highest temperature appears in the direction of 3 points, and the lowest temperature appears in the direction of 0 points. It is suggested that the heating effect should be improved as much as possible, and the heating coil should be tightly wrapped, especially at 6 o ' clock. Building windbreaks when necessary.
(3) The preheating temperature field of the in-service pipeline is related to the gas flow in the pipeline. Under the same heating power, the downstream temperature of the tropical zone is higher than the upstream temperature at the same position, and the closer to the tropical zone, the higher the temperature. When the gas flow rate reaches 9.37m/s and the heating power is 160kW, the measured average temperature of Φ1016 pipeline with 50mm downstream of the tropical zone is 107°C, and the measured average temperature of 50mm upstream of the tropical zone is 71°C. The preheating temperature basically meets the requirements of the current site (60°C). When the gas flow rate reaches 8.91m/s and the heating power is 200 kilowatts, the measured average temperature of Φ1219 pipeline at 50mm downstream of the tropical zone is 72°C, the measured average temperature at 50mm upstream of the tropical zone is 52°C, and the measured average temperature at 30mm upstream of the tropical zone is 71°C. It is. It is suggested that the B-type sleeve should be installed under this condition, and the methods of simultaneous heating of the upstream and downstream, moving the tropics, and auxiliary heating should be used to meet the requirements of welding preheating.

References

1. WU Linjun. Reliability Analysis of Repair Technology of Pressure Pipeline in Service [J]. Quality and Technical Supervision Research, 2020(06):33-36.
2. HAO Jianbin, WU Xinjuan et al. In-Service Pipeline Repair and Simulation Testing Method of Emergency repair and Welding Process of Pipeline. OGST, 2006(02):37-40+62+2.
3. XU Minxu, LUO Lihui, ZHOU Huiping, et al. Influence of B -type Sleeve Welding on the Performance of X80 Pipeline Steel [J]. Materials Protection, 2019,52(06):149-156.
4. Li Rongguang, et al: Improvement on Remediation Technology for Type B Full-Encirclement Steel Sleeves. Oil & Gas Storage and Transportation, 2010,29(10):755-758+717.
5. CHEN Juan, JI Feng, LIU Xuebin, et al. Control measures of in-service pipeline welding hydrogen-induced crack under low temperature environment [J]. Oil & Gas Storage and Transportation, 2014, 33(12):1297-1300.
6. FU Wei, LYU Yuling. Burn-though and Hydrogen Induced Cracking Prediction of Oil and Gas Pipeline during In-service Welding [J]. Hot Working Technology,2018,47(23):241-245.
7. LI Weiwei, HE Xiaodong, GE Jialin. Enlightenment and Reference of Foreign Advanced Standards for Girth Welding of Oil and Gas Pipelines [J]. Petroleum Tubular Goods & Instruments, 2020, 6(02): 1-7.
8. LI Weiwei, MA Qiurong, ZHAO Xinwei, et al. Effect of Pre-heating Temperature at Welding Heat Affected Zone of X80 PipelineSteel [J]. Petroleum Engineering Construction, 2005(04):10-12+4.
9. HE Xiaodong, WANG Yuexia, WU Shuang, et al.. Effect of Pre-heat Temperature and Cooling Rate on Properties in Coarse
10. Grain Heat-affected Zone for X80 High Deformability Pipeline Steel [J]. Welded Pipe and Tube,2012,35(10):10-15.
11. XU Yan, LIU Ying-lai, FENG Zhen-jun, et al.. Effect of Heat Treatment Temperature on Mechanical Tensile Properties of X80 Pipeline Steel [J]. Heat Treatment Technology and Equipment ,2020, 41(06):37-41.
12. Liu Yinglai, Wang Gaofeng, Nie Xianghui, et al..Influence of Normalizing Temperature on Microstructure and Mechanical Properties of Controlled Rolling Steel for Preparation of Manifolds [J]. Natural Gas and Oil, 2015,33(

MATEC Web of Conferences 353, 01002 (2021)
AMME 2021
https://doi.org/10.1051/matecconf/202135301002

Cause analysis of vibration in shell-and-tube heat exchanger

Jia Wu [1,2], Jiansheng Hu[1,2], Bin Du[1,2,*], Ping Tang[1,2] and Jie Tang[1,2]

[1] Zhejiang Academy of Special Equipment Science, Hangzhou, China

[2] Key Laboratory of Special Equipment Safety Testing Technology of Zhejiang Province, Hangzhou, China

Abstract. Heat exchanger is a universal process equipment, which is widely used in chemical industry, oil refining, thermal power and other industries. Shell-and-tube heat exchangers often fail prematurely due to flow-induced vibration. In this paper, aiming at the problem of severe vibration of shell-and-tube heat exchanger in an enterprise, the vibration acceleration sensors are used to measure the vibration of shell and tube side of heat exchanger. Through the analysis of the measurement results, it is found that when the flow rate on the shell side of the heat exchanger is greater than 50000 Nm^3/h, the flow-induced vibration of the tube bundle occurs, and the vibration frequency is 57Hz, which is close to the natural frequency of the tube bundle. And the greater the flow rate on the shell side, the more intense the vibration. In order to reduce the vibration of heat exchanger, the damping measure of inserting baffles between tubes is adopted.

1 Introduction

Shell-and-tube heat exchanger is a widely used process equipment in petroleum, chemical, food and pharmaceutical industries. With the development of social economy and the expansion of industrial production scale, shell-and-tube heat exchangers tend to be large-scale. Moreover, due to the increase of heat exchanger size and tube bundle support spacing, the increase of fluid flow rate, unstable operating conditions and other factors, flow-induced vibration of heat exchanger tube bundle is often caused, resulting in local failure or even overall scrapping of heat exchanger, bring huge economic losses to the factories. There are three main reasons for tube bundle vibration induced by fluid in shell-and-tube heat exchanger, namely vortex shedding, fluid-elastic instability and turbulent buffeting [1].

When the fluid of the shell side flows through the tubes of the heat exchanger, an alternating Karman vortex street will be formed downstream of the tubes, resulting in a periodically varying exciting force, which is perpendicular to the fluid flow direction and leads to tube vibration. When the vortex shedding frequency is consistent with the natural frequency of the tube, violent vibration will occur. In the past decades, many scholars have been studying the vortex shedding phenomenon. Grotz and Rarnold [2] studied the vortex shedding phenomenon in shell-and-tube heat exchangers, proposed a set of design factors, and revealed the relationship between tube spacing ratio and vortex shedding frequency. Lienhard [3] found that for the tube bundle with vortex shedding, the Strouhal number is a function of the tube arrangement. Pettigrew and Taylor [4] formulated design guidelines for prevention of tube bundle failure caused by excessive flow-induced vibration in shell-and-tube heat exchangers. Khushnood and Nizam [5] analyzed the vibration response of tube bundle under two different flow conditions, and studied the influence of damping on tube vibration behavior. Ji et al. [6, 7] discussed the influence of fluid velocity and structural parameters of shell side on flow-induced vibration response of elastic tube bundle.

Fluid-elastic instability is also one of the main causes of heat exchanger tube failure. When the fluid flows through the tube bundle, due to the action of turbulence, some tubes in the tube bundle may move instantaneously, so as to change the flow field around them, destroy the force balance on adjacent tubes, make the tubes move and begin to vibrate. If the amplitude continues to increase, fluid-elastic instability will occur [8]. Roberts [9] found that the displacement mechanism of the tube can be explained as dynamic instability of the tube bundle in the cross flow, and experimentally determined that the alternating motion of the tube will change the wake pairing downstream of the tube bundle. Chen [10] summarized different instability models and stability diagrams to avoid fluid-elastic instability. Fricker [11] found that the clearance between the tube and the support is prone to fluid-elastic instability, so as to predict the vibration characteristics and impact behavior of the tube. Gordon and Lebret [12] analyzed the influence of clearance on fluid-elastic instability through experiments. According to the unsteady flow theory, Cai et al. [13] established the mathematical model of fluid-elastic instability of loosely supported tube in non-uniform cross flow, simulated and studied the characteristics of fluid-elastic instability, and compared the analysis results with the published experimental data. Weaver and Parrondo [14] conducted experimental study on the fluid-elastic instability of tube bundle of multi span heat exchangers, and conservatively estimated the critical velocities of

* Corresponding author e-mail: wujiachn@126.com

seven different configurations according to the semi empirical model and the observed critical instability model. Lin and Yu [15] discussed the problem of fluid-elastic instability by taking multiple cylinders under cross flow as the research object. The results show that when the velocity is higher than the critical value, high amplitude fluid-elastic vibration will occur in multiple cylinders; Above the critical velocity, the cylinders begin to vibrate along the elliptical orbit. Bao et al. [16] theoretically deduced the fluid-elastic excitation on the shell side of the heat exchanger, and analyzed the vibration of the tube bundle at different positions by numerical simulation method.

When the fluid force is the only function of time, it will produce forced vibration, that is, turbulent buffeting. Turbulence is characterized by the random variation of flow pattern and fluid velocity around the tube [17, 18]. De Paula et al. [19] discussed some characteristics of turbulent flow in triangular arranged tubes through hot wire measurement in aerodynamic channels and flow visualization in waterways. Ishihara [20] conducted experiments with actual heat exchangers, measured the vibration of a large number of tubes under different flow rates, and measured natural frequencies and the damping ratio of the tubes. The results show that the cause of tube vibration is the buffeting caused by upstream turbulence. These tubes are supported by many baffles with gaps, which makes the vibration system nonlinear. Goyder [21] pointed out that compared with fluid-elastic instability, the vibration caused by turbulent buffeting is very small. However, after many years of use, due to fretting wear, this relentless low-level vibration will lead to the thinning of the tube wall, which is unacceptable in high-integrity heat exchangers.

In this paper, aiming at the problem of severe vibration of shell-and-tube heat exchanger in an enterprise, the vibration acceleration sensors are used to measure the vibration of shell and tube side of heat exchanger. By analyzing the vibration characteristics, the cause of vibration of heat exchanger is found out, and the vibration reduction measure is put forward.

2 Measuring method

2.1 Heat exchanger model

Flue gas heat exchanger is used to heat the flue gas treated by the desulfurization system to raise the temperature before entering the chimney and improve the corrosion of the tail flue and chimney. The model of a flue gas heat exchanger system of an enterprise is shown in Figure 1. The type of the heat exchanger is shell-and-tube heat exchanger, which uses the capacity of induced draft fan to guide the flue gas.

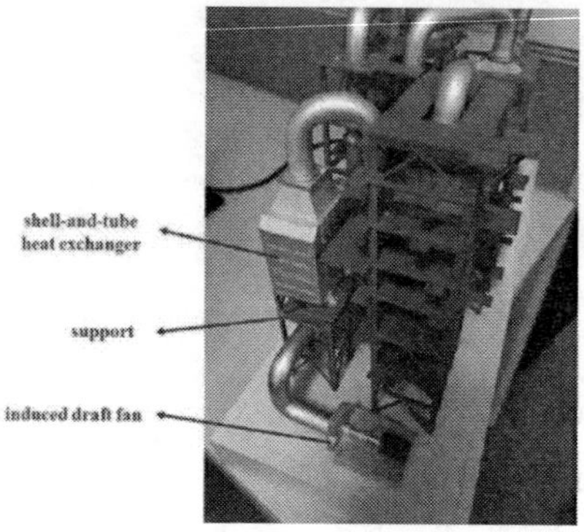

Figure 1. Model of flue gas heat exchanger system

In the shell-and-tube heat exchanger, the stroke of the fluid for heat exchange in the tube bundle is called tube side, and the stroke of the fluid for heat exchange outside the tube bundle is called shell side. The wall of the tube bundle is the heat transfer surface of cold and hot fluid. The tube side and shell side of shell-and-tube heat exchanger are shown in Figure 2, and the process parameters are listed in Table 1.

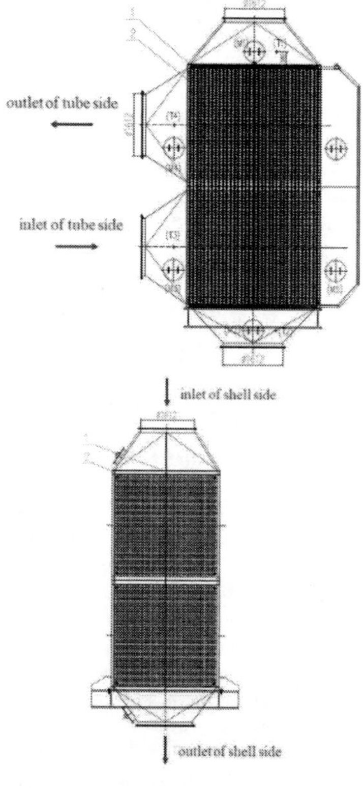

Figure 2. Tube side and shell side

Table 1. Process parameters of shell-and-tube heat exchanger

	Shell side	Tube side
Inlet / outlet temperature	180℃/145℃	115℃/150℃
Pressure drop	1100 Pa	700 Pa
Design flow rate of flue gas	72000 Nm^3/h	72000 Nm^3/h
Flue gas velocity	6.4 m/s	12.4 m/s
Thermal load	0.99 MW	

2.2 Pulse excitation method

The measurement of natural frequencies of tube side and shell side in shell-and-tube heat exchanger is carried out by pulse excitation method, and its principle is as follows:

The function δ represents an instantaneous pulse and if the force acting on the object is a unit impulse $\delta(t)$, then:

$$\delta(t)=\begin{cases}\infty & t=0\\ 0 & t>0\end{cases} \text{ and } \int_{-\infty}^{+\infty}\delta(t)dt=1 \quad (1)$$

In a linear system, assuming that the input signal is $x(t)$, the mathematical operation of the system on the input signal is represented by the symbol L, and the output of the system is $y(t)$, then:

$$y(t)=L[x(t)] \quad (2)$$

In order to obtain the frequency domain characteristics of the system, Fourier transform is performed on $x(t)$, $y(t)$ and system impulse response $h(t)$, and the results are represented by symbols $X(f)$, $Y(f)$ and $H(f)$, respectively, then:

$$\begin{aligned}Y(f)&=\int_{-\infty}^{+\infty}y(t)e^{-j2\pi ft}dt\\&=\int_{-\infty}^{+\infty}\left[\int_{-\infty}^{\infty}x(\tau)h(t-\tau)d\tau\right]e^{-j2\pi ft}dt\\&=\int_{-\infty}^{+\infty}x(\tau)\left[\int_{-\infty}^{\infty}h(t-\tau)e^{-j2\pi f(t-\tau)}dt\right]e^{-j2\pi ft}d\tau\\&=\int_{-\infty}^{+\infty}x(\tau)H(f)e^{-j2\pi f\tau}d\tau\\&=X(f)H(f)\end{aligned} \quad (3)$$

Because when the input signal $x(t)$ is equal to $\delta(t)$, its Fourier transform $X(f)$ is equal to 1. It can be obtained from the above formula:

$$H(f)=\frac{Y(f)}{X(f)}=\int_{-\infty}^{+\infty}y(t)e^{-j2\pi ft}dt \quad (4)$$

This shows that if a unit impulse force is applied to the measured system, as long as the response of the system to the impulse force is measured and Fourier transformed, the frequency response function of the system can be obtained, which represents the dynamic characteristics of the system.

3 Measurement results

3.1 Natural frequency measurement

The natural frequency measurement is carried out by pulse excitation method. By analyzing the frequency spectrum of pulse response, the natural frequency of tube side and shell side can be obtained.

Figure 3 shows the vibration spectra of the tube side without excitation and excitation, and Figure 4 shows the vibration spectra of the shell side without excitation and excitation.

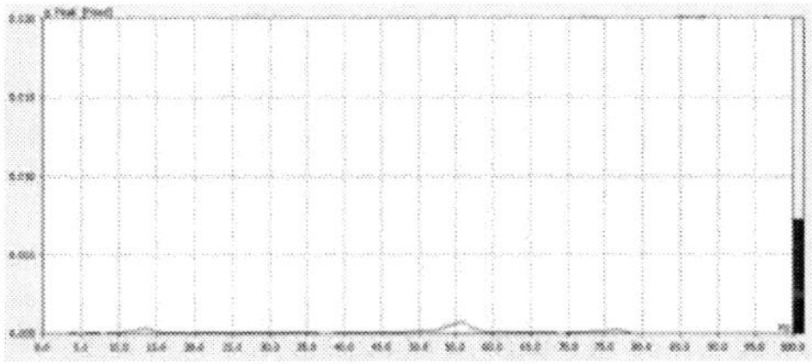

(a) without excitation

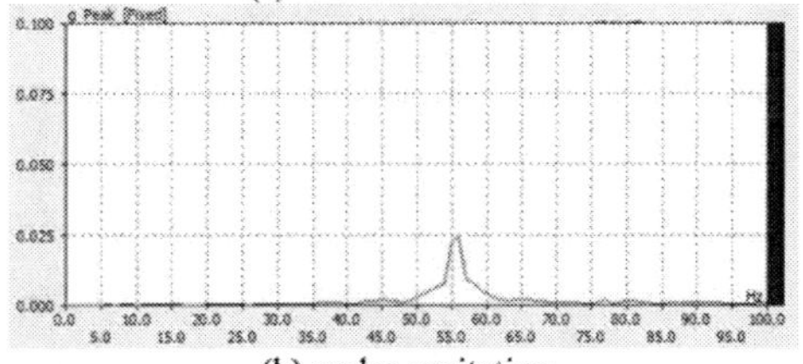

(b) under excitation

Figure 3. Vibration spectra of tube side without excitation and excitation

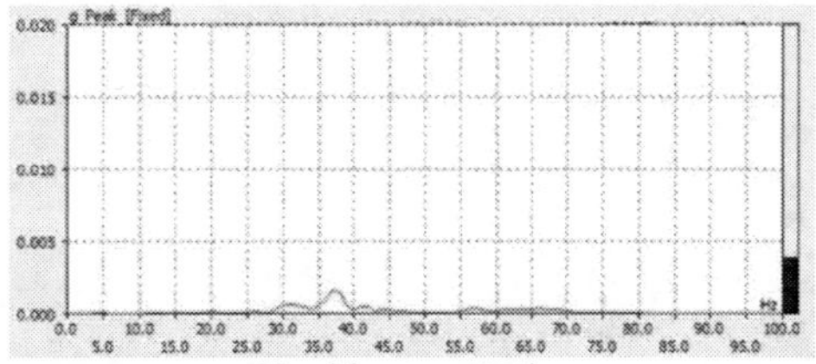

(a) without excitation

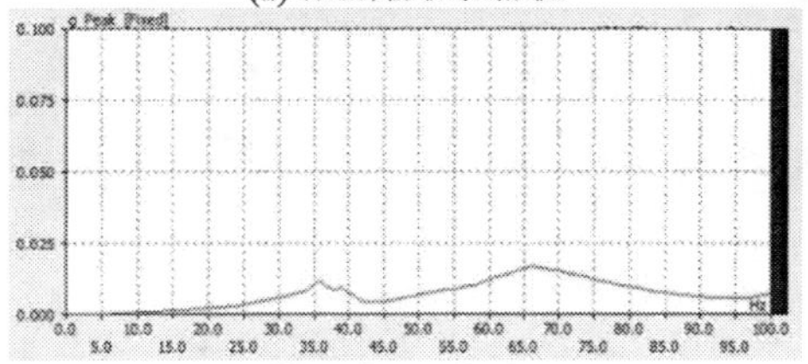

(b) under excitation

Figure 4. Vibration spectra of shell side without excitation and excitation

As can be seen from Figure 3, when not excited, the frequency of the tube side is mainly 56Hz, while when excited, the frequency still exists, but the amplitude increases sharply. This indicates that the natural frequency of the tube side is 56Hz. Similarly, it can be seen from Figure 4 that the natural frequencies of the shell side are 37Hz and 66Hz.

3.2 Comparison of vibration under different flow rates

Figure 5~9 shows the vibration of shell-and-tube heat exchanger under five different flow rates. It can be seen from the figures that when the flow rate is 45000 Nm^3/h and 48000 Nm^3/h, the vibration is small, and the main vibration frequency is about 37.5Hz, which is very close to the natural frequency of 37Hz on the shell side. When the flow rate is greater than 50000 Nm^3/h, the vibration increases significantly, and the main vibration frequency is about 57Hz, which is close to the natural frequency of the tube side. It shows that when the flow rate increases to a certain extent, flow-induced vibration occurs in the shell-and-tube heat exchanger.

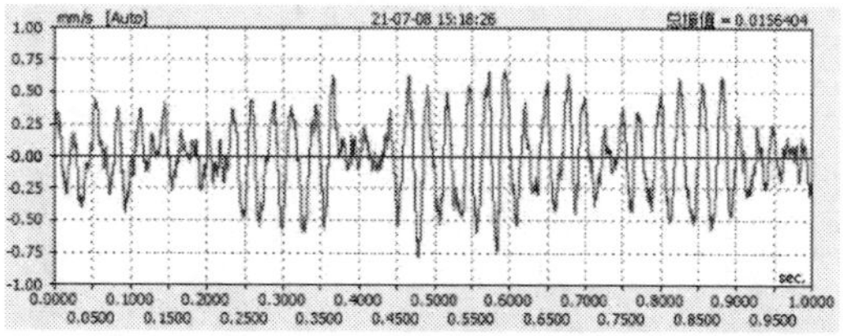

(a) vibration velocity waveform

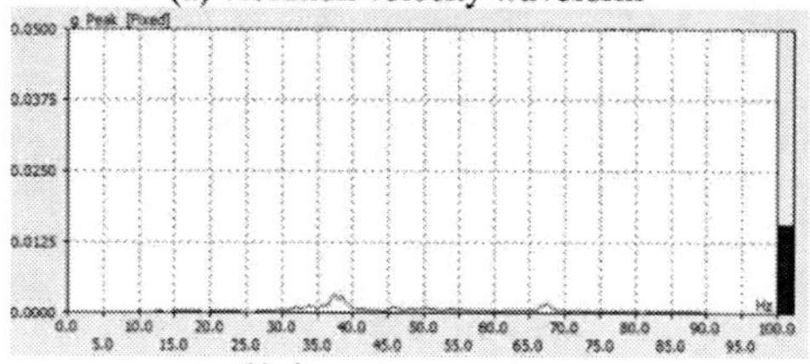

(b) frequency spectrum

Figure 5. Vibration when the flow rate is 45000 Nm^3/h

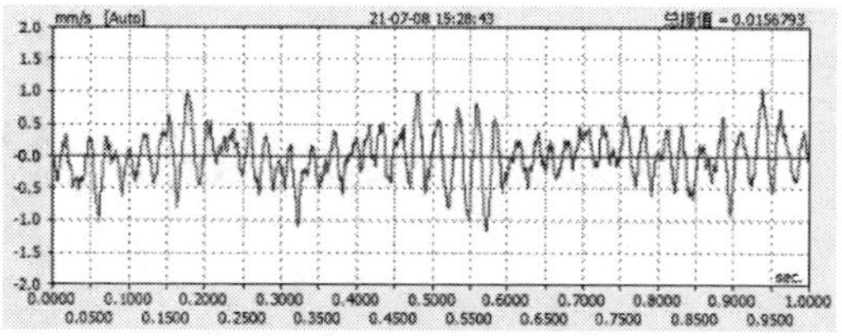

(a) vibration velocity waveform

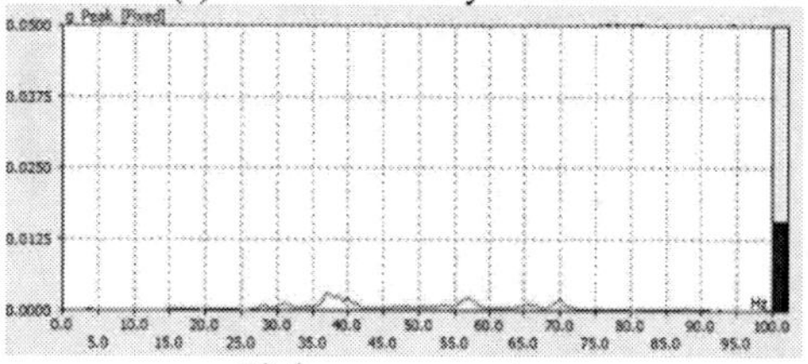

(b) frequency spectrum

Figure 6. Vibration when the flow rate is 48000 Nm^3/h

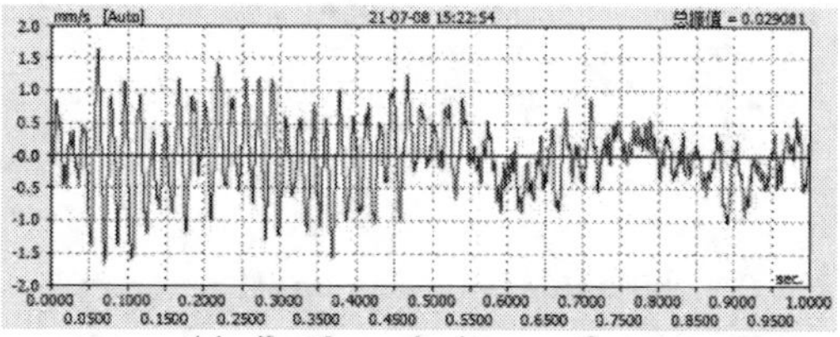

(a) vibration velocity waveform

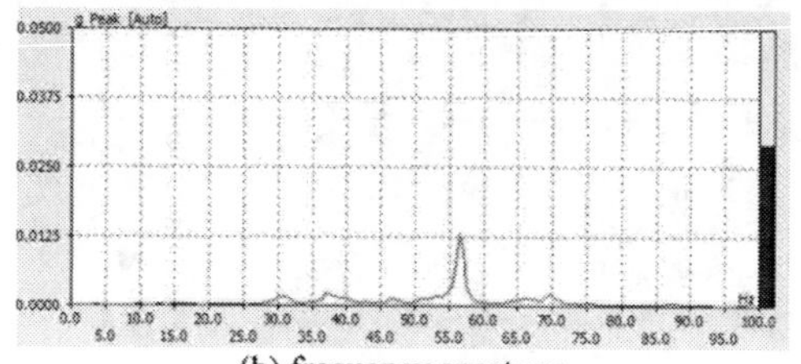

(b) frequency spectrum

Figure 7. Vibration when the flow rate is 50000 Nm^3/h

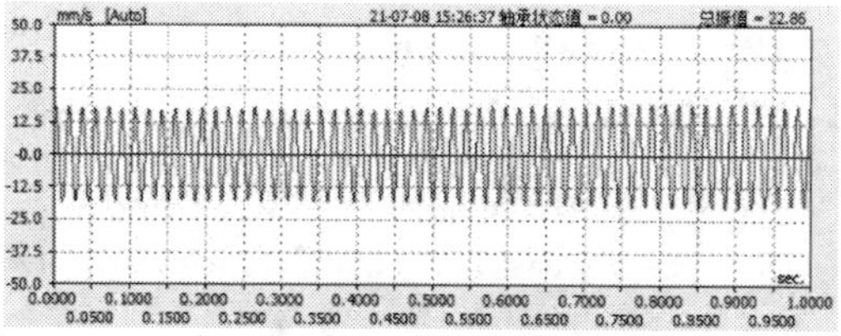

(a) vibration velocity waveform

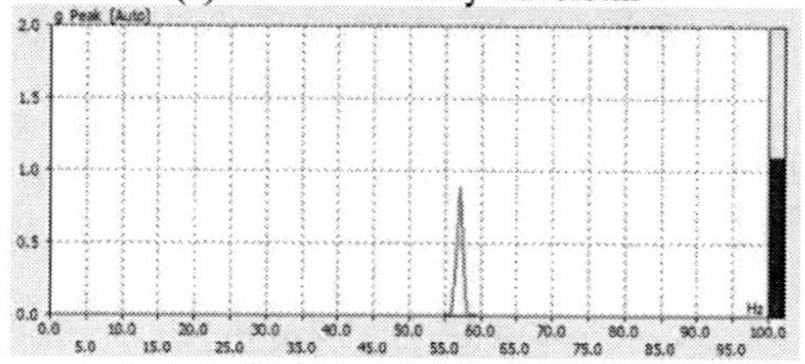

(b) frequency spectrum

Figure 8. Vibration when the flow rate is 51000 Nm^3/h

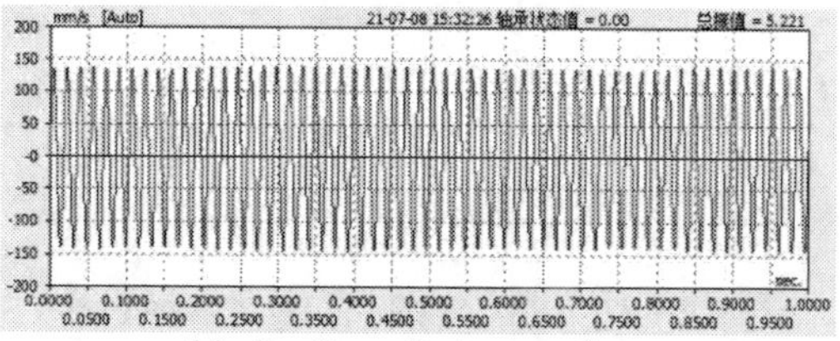

(a) vibration velocity waveform

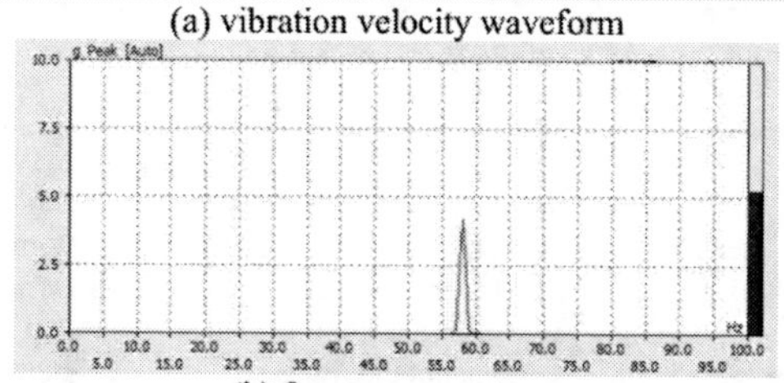

(b) frequency spectrum

Figure 9. Vibration when the flow rate is 55000 Nm^3/h

4 Vibration reduction measure

The vibration of the tube in heat exchanger is inevitable. Slight vibration will not cause damage to the heat exchanger, but also strengthen heat transfer and reduce scaling. However, for strong vibration, necessary anti vibration measures shall be taken to slow down the vibration and avoid vibration damage of heat exchanger. According to the above analysis, the fundamental way to reduce vibration is to avoid the natural frequency of the tube as far as possible.

Because the natural frequency of the heat exchange tube is inversely proportional to the square of the supporting span of the tube bundle, reducing the span can improve the natural frequency of the heat exchange tube, which is the most economical and effective vibration reduction measure in engineering practice. In Figure 10, two baffles are inserted between the heat exchange tubes, which reduces the span, strengthens the tube bundle and changes the natural frequency of the heat exchange tube.

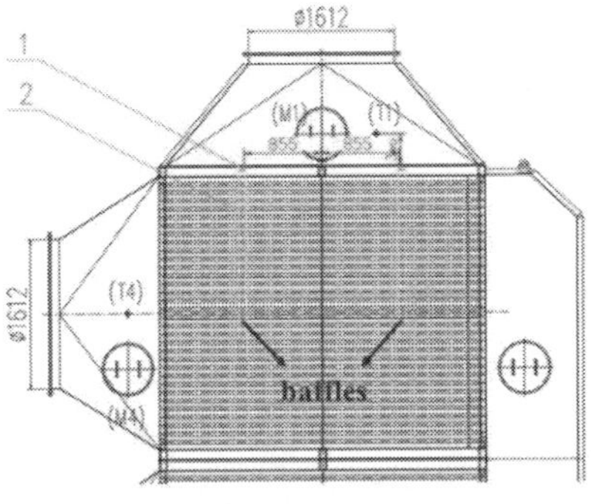

Figure 10. Baffles inserted between heat exchange tubes

5 Conclusions

(1) The vibration of the shell-and-tube heat exchanger is caused by flow-induced vibration of tube bundle. When the flow rate on the shell side of the heat exchanger is greater than 50000 Nm^3/h, flow-induced vibration is generated, and the vibration frequency is 57Hz, which is close to the natural frequency of the tube bundle. And the greater the flow rate on the shell side, the greater the impact on the tube bundle and the more intense the vibration.

(2) In order to reduce the vibration of the heat exchanger, it is suggested to insert baffles between the heat exchange tubes, so as to change the natural frequency of the heat exchange tubes and avoid the frequency of exciting force.

Acknowledgments

This work was financially supported by the Natural Science Foundation of Zhejiang Province (NO. LQ21E050001) and the science and technology project of Zhejiang Provincial Administration for Market Regulation (NO. CY2022221).

References

1. Weaver, D. T., and Fitzpatrick, J. A. (1988). A review of cross-flow induced vibrations in heat exchanger tube arrays. Journal of fluids and structures, 2(1), 73-93.
2. Grotz, B. J., and Rarnold, F. (1956). Flow-induced vibrations in Heat Exchangers. Stanford University, Department of Mechanical Engineering.
3. Lienhard, J. H. (1966). Synopsis of lift, drag, and vortex frequency data for rigid circular cylinders (Vol. 300). Pullman, WA: Technical Extension Service, Washington State University.
4. Pettigrew, M. J., and Taylor, C. E. (2003). Vibration analysis of shell-and-tube heat exchangers: an overview—Part 2: vibration response, fretting-wear, guidelines. Journal of Fluids and Structures, 18(5), 485-500.
5. Khushnood, S., and Nizam, L. A. (2017). Experimental study on cross-flow induced vibrations in heat exchanger tube bundle. China Ocean Engineering, 31(1), 91-97.
6. Ji, J., Ge, P., and Bi, W. (2016). Numerical analysis on shell-side flow-induced vibration and heat transfer characteristics of elastic tube bundle in heat exchanger. Applied Thermal Engineering, 107, 544-551.
7. Ji, J. D., Ge, P. Q., and Bi, W. B. (2018). Numerical analysis of shell-side flow-induced vibration of elastic tube bundle in heat exchanger. Journal of Hydrodynamics, 30(2), 249-257.
8. Price, S. J. (1995). A review of theoretical models for fluidelastic instability of cylinder arrays in cross-flow. Journal of Fluids and Structures, 9(5), 463-518.
9. Roberts, B. W. (1966). Low frequency, aeroelastic vibrations in a cascade of circular cylinders. Mechanical Science Monograph, (4).
10. Chen, S. S. (1984). Guidelines for the instability flow velocity of tube arrays in crossflow. Journal of Sound and Vibration, 93(3), 439-455.
11. Fricker, A. J. (1992). Numerical analysis of the fluidelastic vibration of a steam generator tube with loose supports. Journal of Fluids and Structures, 6(1), 85-107.
12. Godon, J. L., and Lebret, J. (1988). Influence of the tube-support plate clearance on flow-induced vibration in large condensers. In Proceedings of the International Symposium on Flow-Induced Vibration and Noise: Flow-Induced Vibration in Heat-Transfer Equipment, Chicago, IL (pp. 177-186).
13. Cai, Y., Chen, S. S., and Chandra, S. (1992). A theory for fluidelastic instability of tube-support-plate-inactive modes. ASME. J. Pressure Vessel Technol, 114(2): 139–148.
14. Weaver, D. S., and Parrondo, J. (1991). Fluidelastic instability in multispan heat exchanger tube arrays. Journal of fluids and structures, 5(3), 323-338.
15. Lin, T. K., and Yu, M. H. (2005). An experimental study on the cross-flow vibration of a flexible cylinder in cylinder arrays. Experimental thermal and fluid science, 29(4), 523-536.
16. Bao, M., Wang, L., Li, W., and Gao, T. (2017). The Vibration Analysis of Tube Bundles Induced by Fluid Elastic Excitation in Shell Side of Heat Exchanger. In

IOP Conference Series: Materials Science and Engineering (Vol. 239, No. 1, p. 012011). IOP Publishing.

17. Blevins, R. D. (1977). Flow-induced vibration. New York.
18. Polak, D. R., and Weaver, D. S. (1995). Vortex shedding in normal triangular tube arrays. Journal of Fluids and Structures, 9(1), 1-17.
19. De Paula, A. V., Endres, L. A. M., and Möller, S. V. (2012). Bistable features of the turbulent flow in tube banks of triangular arrangement. Nuclear engineering and design, 249, 379-387.
20. Ishihara, K. (2007). Flow Induced Vibration of Shell & Tube Type Heat Exchanger (1st Report, Understanding of Phenomena). Journal of Environment and Engineering, 2(2), 293-302.
21. Goyder, H. G. D. (2002). Flow-induced vibration in heat exchangers. Chemical Engineering Research and Design, 80(3), 226-232.

https://doi.org/10.1051/matecconf/202135301003

Stress Calculations Adopted Finite Element Method of Pressure Steam Pipeline Containing Defects

Yonggui Chen [1,2], Ping Tang [1,2,*], Jiang Zhong [3], Sunting Yan [1,2], Nanhui Jin [1,2], Lintao Liu [3]

[1] Zhejiang Academy of Special Equipment Science, Hangzhou 310020, China
[2] Key Laboratory of Special Equipment Safety Testing Technology of Zhejiang Province, Hangzhou 310020, China
[3] Tongxiang Ties Environmental Energy co., Ltd., China

Abstract. The stress behavior of defects is the key factor in the safety assessment of pressure steam pipeline containing defects. For achieving carbon peaking and carbon neutrality, high-parameter pressure steam pipelines are widely designed in the recent years. It makes the safety assessment based on defect stresses become significantly important. A combined FEA method of stress calculations was proposed in the present study for calculating film stress and bend stress of defects located in a pressure steam pipeline from a certain power plant. An efficient program code with graphical user interface was designed to automatically generate FEA models of defect part pipelines. It is truly an innovative highlight of the present work attributed to convenient operation and outstanding efficiency. This paper provides an available way to obtain the stress state of defects in the safety assessment of pressure steam pipeline containing defects.

Key Words. Pressure steam pipeline, Large-scale defects, Stress state, Structural integrity.

1 Introduction

Pressure steam pipeline is general equipment in petrochemical and power industry. The failure of its structural integrity may result in costly financial losss, severe environmental harms and catastrophic accidents of many casualties[1]. With the development trend of achieving carbon peaking and carbon neutrality through energy saving, emission reduction, and improvement of energy efficiency ratio in various countries around the world, an amount of high-parameter steam pipeline are widely designed, whose safe operateion is mostly threatened by pipeline defects associated with mechanical damage. Such high-temperature and high-pressure working environment may more rapidly lead to plastic failure of pipeline defects [2-3]. Therefore, the safety assessment of pressure steam pipeline containing defects based on actual defect stresses has become an increasingly urgent issue[4-6].

However, it is not easy to get the magnitude and distribution of defect stresses of pressure pipelines. Experimental measurement is hard and costly although as the mostly direct way. The complicated factors including pipeline structure, pipeline layout, support constraints and external loads determine that accurate theoretical calculations of defect stresses are difficult to accomplish[7]. Besides, The huge capacity of 3D shell element models presents inefficient computing power of the used numerical iterative calculations. Given the above, this paper introduces a Finite Element Analysis (FEA) process that is conducted to calculate the defect stresses of a boiler pressure steam pipeline from a certain power plant. This method combines the overall and partial pipeline modeling, which provides a new way of stress acquisition during the safety assessment of pressure steam pipeline containing defects.

2 Structure Presentation of Pressure steam pipeline

In a statutory periodical inspection process for a certain power plant, the boiler pressure steam pipeline as shown in Fig.1 was detected with two non-fusion defects of defect ZZQ-113 and PZQ-8, which laid a safety hazard for the normal operation of the pipeline. The working temperature and pressure of the pipeline are 400℃ and 4.2MPa. According to the safety status level of the standard of TSG 21-2016"Supervision Regulation on Safety Technology for Stationary Pressure Vessel", two Defects are assessed as grade IV. In order to judge whether the pipeline can operate safely in the next statutory period, it is necessary to assess the fitness-for-service of the pipeline according to GB/T 19624-2019 "Safety Assessment of In-service Pressure Vessels Containing Defects". Forces and stress distributions at section position of pipeline defects, revealing the safe state of defects, are hard and key points in the safety assessment processes. Following study takes the stress

* Corresponding author: tangpingjob@163.com

value of defect ZZQ-113 and PZQ-8 as the destination of calculations, aiming to describe universal finite element calculation method of defect stresses in pressure steam pipeline containing defects.

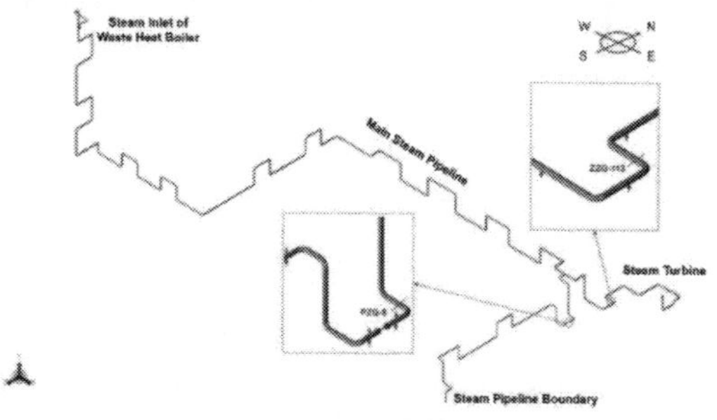

Figure 1. Pressure Steam Pipeline Geometry

3 Solution Scheme of Pipeline Stress

Given the difficulty of accomplishment of experimental measurement and theoretical calculation, the Finite Element Method (FEM) has become an available and effective way to obtain required data for researchers. There are two obvious shortcomings when used 3D FEM analysis ways are directly applied to solve defect stresses of pressure steam pipeline. The 3D FEM analysis model of a substantive pressure pipeline is full of a great number of grids that is inefficient for numerical iterative calculations, with higher CPU performance requirements. In addition, it is hard to involve complex factors such as rigid support load and self-adaptive spring hanger load of in-service pressure pipeline for common 3D FEA sofewares. CAESAR II developed by COADE company in the United States is the most widely used software of stress analysis and check for pressure pipeline in various industries. This tool fully takes into account the complex factors described above of in-service pressure pipeline. Its 1D beam element model consisted of simple nodes greatly improves the efficiency of iterative calculations. However, the stress data based on 1D beam element model is the section-averaged value in the node section location, which cannot express the most serious stress state in which defects are sensitive and vulnerable. Therefore, node stress solved by CAESAR II sofeware is powerless for the safety assessment of pressure pipeline containing defects.

Here, a method combines the overall and partial pipeline modeling is introduced for stress calculations of defects. There are two major contents of solution scheme. Firstly, it is needed to build pipline-wide 1D beam element CAESAR II calculation model of in-service assessed pressure pipeline in particular to set rigid support and self-adaptive spring hanger. The model makes full use of the advantages of CAESAR II software in modeling and calculating aspects, Comprehensively considering pipeline fittings, support constraints and external loads. A key point in this step is to empirically select and set the segmentation nodes based on locations of detected large-scale defects in the established pipline-wide model as the boundary of local defect part model. In iterative solutions, the force and moment of each node are calculated. The second content is to invoke 3D FEA commercial software to establish the shell element models of the local defect parts. Boundary conditions are imposed on these models equivalently from the above calculations described. Goal stress distributions of defects are presented by performing numerical iterative calculations by FEA solver where the principal stresses of defects are calculated based on the plastic strength level and the linear elastic theory of the material, with the assumption that there are no defects at detected defect locations of the pipeline. The solution process of goal stress is shown in Fig.2.

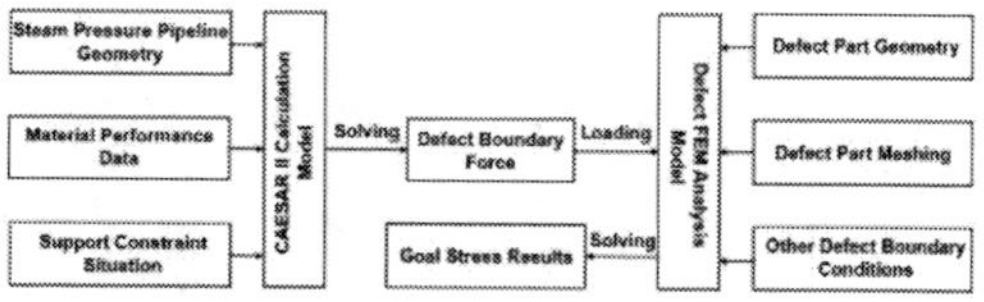

Figure 2. Defect Stress Solution Flowchart

Usually, the pipeline stress is non-linearly distributed over the thickness of pipeline. Goal stresses including the film stress σ_m and the bending stress σ_B used to assess pressure pipeline safety is the results of decomposing and calculating σ_1 and σ_2 according to formula (1). There σ_1 and σ_2 are the stress magnitudes on the inner and outer walls after stress linearization on the cross section of the defect, respectively, MPa. The stress linearization method of common surface defects is shown in Fig.3.

$$\begin{cases} \sigma_m = (\sigma_1 + \sigma_2)/2 \\ \sigma_B = (\sigma_1 - \sigma_2)/2 \end{cases} \tag{1}$$

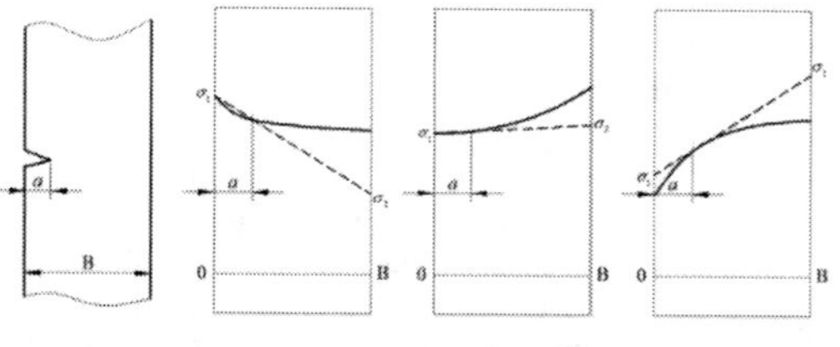

Figure 3. Stress Linearization Method of Surface Defects

4 Efficient Program Code Design

In some defect safety assessments of pressure steam pipelines containing defects, a number of FEA models of defect part pipeline may be needed to build, rather than only two defects in the current pressure steam pipeline. However, such work is repetitive, laborious and time-consuming. And it does not mean that is simple or easy work. The modeling and meshing of 3D solid pipeline are very technical and highly experienced operation, especially for T-tubes, elbows and variable diameter tubes. The quality of meshing determines the accuracy and reliability of stress calculation results, which is the important guarantee for assessing pipeline safety. In order to save the time of building defect part models and ensure the grid quality of meshing defect part models, a program code in python language is designed. This code describes 3D graphics of pressure steam pipeline into computer language based on the mathematical algorithms of computer graphics, and realizes the stereo display of pressure steam pipeline in the user graphical interface. Then according to geometric topology, how to perfectly mesh pipeline model is conceived and accomplish by compiled code. Eventually, the building and meshing process of defect part models is automatically realized. Besides, the program code also has the function of imposing all boundary conditions including various types of support load. The graphical user interface developed is shown in Fig.4.

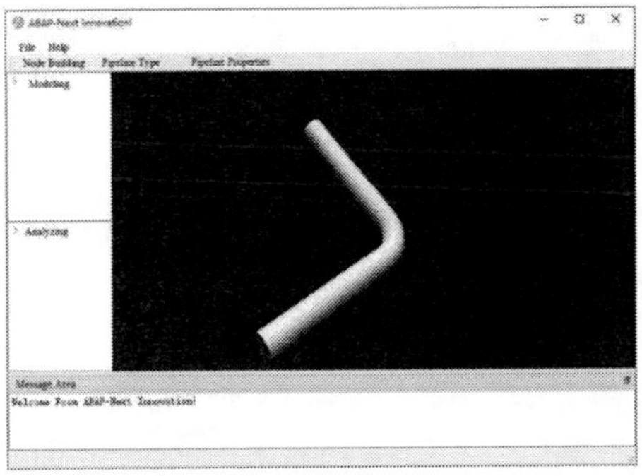

Figure 4. Graphical User Interface of Designed Program Code

All implementations of defect solution are operated in the graphical user interface. General input has following contents: the type of various pipeline segment, the coordinates of geometry nodes of line diagram, the outer diameter and the wall thickness , and so forth. Besides, the material and media properties of the pipeline and all the force boundary conditions are also necessary. Once the analyzing instruction is submitted, the designed program code directly and quickly invokes the FEA software installed on the computer to generate the 3D defect part model, to mesh the finite element model, to load the boundary conditions and to perform mechanical calculations in rapid sequence. The whole process runs automatically. The program code efficiently conducts complicated modeling and meshing of defect part geometry by a labor-saving way that just one-line nodes of pipelines are defined. All the formed grids are hexahedral structured grids with superior computational performance. The modeling and meshing process of the pressure steam pipeline defect ZZQ-113 is shown in Fig.5. With remarkable results, hours of work of defect analyses was completed in minutes with the designed efficient program code. Program code design above is truly an innovative highlight of the present stress calculation method.

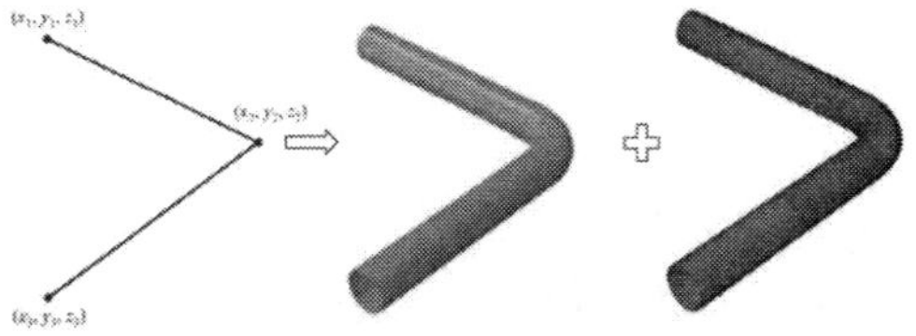

Figure 5. Efficient Modeling and meshing of Defect ZZQ-113

5 Stress Calculation Results

After the analyzing instruction was submitted, as results of iterative solutions, the stress distributions of defect part pipelines were presented in the form of a color cloud diagram. According to the stress distribution of defect ZZQ-113 as shown in Fig.6, the goal stresses of the defect were calculated through the stress linearization process performed on the cross-section of the defect. A stress path over wall thickness of pipeline was defined by picking start point and end point at which stress state was the most severe. Then the non-linear stress curves of dozens of nodes on this path were obtained. Linearize the stress curves in the way described above at the chapter of solution scheme of pipeline stress. Goal stresses both the film stress σ_m and the bending stress σ_B were calculated by substituding stress σ_1 and σ_2 on the inner and outer walls into formula (1). Finally, the film stress and the bending stress of the defect ZZQ-113 were 59.19 MPa and 8.96 MPa, respectively. With the same treatment, the film stress of the defect PZQ-8 was 60.20 MPa; and the bending stress was 43.79 MPa.

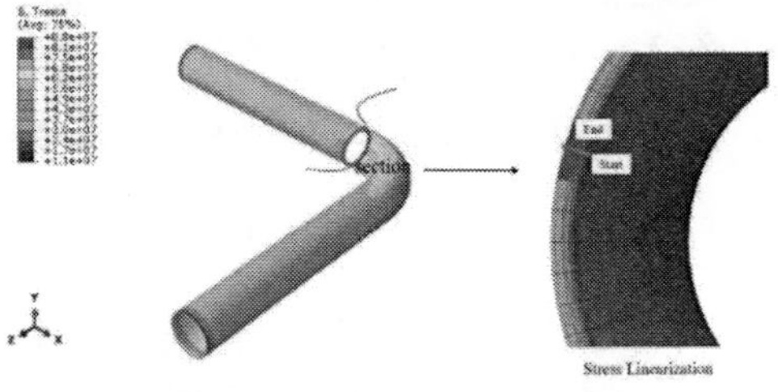

Figure 6. Stress Linearization of Defects ZZQ-113

6 Conclusions

The film stress and bend stress of two defects located on a pressure steam pipeline from a certain power plant were calculated in the present study. From this example, a combined FEA method of stress calculations was proposed. Follow this method, defect stresses are calculated with CAESRA II model and defect FEM analysis model in combination. Thereafter, an efficient program code with graphical user interface was designed to invokes the FEA software to automatically generate the 3D defect part model, to mesh the finite element model, to load the boundary conditions and to perform mechanical calculations in rapid sequence. In the current calculations, hours of work of defect analyses was completed in minutes with the designed code. That's a

 https://doi.org/10.1051/matecconf/202135301003

pleasant presentation of efficiency improvement. This paper provides an available way to obtain the stress state of the safety assessment of pressure steam pipeline containing defects.

Acknowledgments

This work was supported by Natural Science Foundation of Zhejiang Province (LQ21E050001) and the science and technology project of Zhejiang Provincial Administration for Market Regulation (CY2022221).

References

1. Pinheiro B, Soares C G, Pasqualino I. Generalized expressions for stress concentration factors of pipeline plain dents under cyclic internal pressure[J]. International Journal of Pressure Vessels and Piping, 2019, 170:82-91.
2. Witek M. Structural integrity of steel pipeline with clusters of corrosion defects[J]. Materials, 2021, 14(852):1-15.
3. Marti I, Sedmak A, Mitrovic N, et al. Effect of over-pressure on pipeline structural integrity[J]. Tehnicki Vjesnik, 2019, 26(3):852-855.
4. Dotta F, Ruggieri C. Structural integrity assessments of high pressure pipelines with axial flaws using a micromechanics model[J]. International Journal of Pressure Vessels & Piping, 2004, 81(9):761-770.
5. Adib-Ramezani H, Jeong J, Pluvinage G. Structural integrity evaluation of X52 gas pipes subjected to external corrosion defects using the SINTAP procedure[J]. International Journal of Pressure Vessels & Piping, 2006, 83(6):420-432.
6. Kec J, Černý, Ivo, J Luštinec. Experimental verification of dents effect on the structural integrity of pipeline DN 300[J]. Key Engineering Materials, 2018, 774:319-324.
7. Yuanlong Lin, Zhenghong Yan. Application of CAESAR II in safety assessment of pressure pipe[J]. Petro-chemical Equipment, 2017(01):70-73.

Research status of semi-permanent release agent

Bo Liang [1, 2,*], *Zhonghuan Zhang* [2], *Tong Yin* [2], *Shuai Yang* [2], *Junpeng Shao* [2], *William Manchester* [3]

[1] Shandong Meichen Industrial Group Co., LTD., Zhucheng 262200, China
[2] Harbin University of Science and Technology, Harbin 150000, China
[3] Krirk University, Bangkok 10220, Thailand

Abstract. According to the composition of substances, release agent can be divided into organic and inorganic two types, inorganic release agent includes mica powder, graphite powder, etc., organic release agent mainly includes organic silicon, organic fluorin. According to the service life, release agent can be divided into one-time release agent and semi-permanent release agent. This paper introduced several kinds of releasing agent, summarized and analyzed the advantages of semi-permanent release agent and the present situation of domestic and foreign research, and prospected the future of release agent.

Keywords: Release agent, Semi-permanent release agent, Rubber products

1 Introduction

China is a major producer and consumer of plastic and rubber industry. Since 2010, the annual production of plastic and rubber has exceeded 60 million tons, ranking first in the world in both annual production and consumption capacity. With the rise of rubber plastic industry, the amount of release agent is also increasing year by year, so the market size is huge[1-9].

Especially in the automobile hose industry, the molding of automobile hose needs to go through plasticizing, mixing, pressing, calendering and vulcanization several processes. Vulcanization is an important process to make rubber and skeleton layer produce high adhesive strength, so that it becomes a whole, is an important link to ensure the quality of rubber hose. In the process of vulcanization, the rubber hose and the mandrel are easy to adhesion, so before vulcanization, the mandrel needs to be coated with a release agent to prevent adhesion[10-15]. However, there are many problems such as oil bleaching, yellowing, poor stability and less demoulding times in the current rubber hose demoulding agent[1, 16-22].

Due to the effect of high temperature for a long time, the product will form dirt on the mold in the casting process, resulting in the phenomenon of product adhesion, so it is necessary to form a durable protective film on the product to ensure the smooth demoulding. The silicone release agent is an important additive to form this layer of protective film. The organosilicon compound has a variety of practical functions, which has a short film forming time at low temperature, good high temperature resistance, is often used in the demoulding process of a basic raw material[23-25].

Because of the existence of silicon-oxygen bond in its molecular structure, the bond energy and bond angle of organosilicon compound in the organosilicon demoulding agent are larger. Therefore, it has high and low temperature resistance, radiation resistance, oxidation resistance, low surface tension, hydrophobicity, electrical insulation and other characteristics. Because of these properties, organosilicon compounds are widely used in textile industry, food industry and construction industry [26-29].

However, with the improvement of people's living standard and the increase of industry demand, the functional requirements of organic silicon are also gradually increasing[30-36]. At the same time, due to the large hydrophobicity of methyl silicone oil, it is difficult to emulsify and the microemulsion stability is not good, resulting in oil and water separation, oil bleaching, yellowing and other problems in the process of use, so that the demoulding effect becomes poor. In order to meet the requirements of social industry progress, and improve the technical problems of methyl silicone oil release agent instability, it is necessary to reform the structure of organic silicon compound, to obtain more excellent and complete performance.

The rubber products industry in the world is booming and the rubber output is rising steadily. As a big country of rubber production and consumption, the rubber industry has been in a state of rapid development since the 21st century. With the rise of rubber industry, release agent as an indispensable additive rubber products have also been developed, a large number of domestic and foreign research on release agent has been launched. In the process of vulcanization of rubber products, when the products are peeled from the mold, friction resistance may be generated because of the surface defect of the mold,

* Corresponding author: bliang0325@163.com

which increases the difficulty of peeling[37]. Release agent can not only reduce the product and mold peeling force to decrease product demoulding difficulty, but also protect the product and mold surface. It plays an irreplaceable role in the vulcanization of rubber products.

2 Classification of release agent

According to the composition of substances, release agent can be divided into organic and inorganic two types, inorganic release agent includes mica powder, graphite powder, etc., organic release agent mainly includes organic silicon, organic fluorine, wax system, etc[38]. Silicone series and fluorine series are widely used because of their excellent physical and chemical properties.

According to the service life, release agent can be divided into one-time release agent and semi-permanent release agent[37, 39]. The release agent is coated on the surface of the mold before each production cycle, and demoulding is achieved by destroying the cohesiveness of the release agent, which is called the disposable release agent. The main disadvantage of disposable release agent is that when the product is released from the mold, most of the release agent will migrate to the surface of the product resulting in huge loss and decrease in the accuracy of the product, so the disposable release agent is also known as migration release agent or sacrifice release agent.

Semi-permanent release agent is generally made of fluorine and silicon crosslinked polymer as the main raw material. The semi-permanent release agent is applied to the surface of the mold. The solvent volatilizes at room temperature or high temperature, and the solute is crosslinked and solidified, forming a complete film on the surface of the mold. Compared with the disposable release agent, the formation of the cross-linked film greatly improves the release performance. Due to the large adhesion of the cross-linked film, the release agent does not migrate to the product after the product molding demoulding process. so the semi-permanent release agent is also known as the non-migratory release agent. Due to the little consumption of release agent in the production process, it not only reduces the raw material cost but also reduces the manual cleaning mold time, greatly improving the production efficiency. so semi-permanent release agent is widely used, semi-permanent release agent accounts for up to 25% in the rubber industry[37, 40-44].

3 Superiority of semi-permanent release agent

Compared with other types of release agent, semi-permanent release agent will form a chemically stable and heat-resistant crosslinking film on the surface of the mold, because this crosslinking film is difficult to be decomposed, so semi-permanent release agent has the following advantages[40, 42, 44, 45]:

(1) In the demoulding process of vulcanized rubber products, the pollution of the mold caused by the migration of paraffin series release agent is avoided.
(2) The friction coefficient of crosslinked film is very small, which can protect rubber products and molds well and ensure that the surface of molds remains smooth
(3) There is no need to worry about subsequent bond failures.
(4) Because of the stable chemical properties of the crosslinked membrane, there is no need to worry about its reaction with the decomposition of rubber in the high temperature of vulcanization.
(5) There is little consumption before and after product demoulding, high demoulding times, low release agent and labor cost, and high production efficiency.

4 Research status of semi-permanent release agent

With the development of rubber products industry, the research of release agent has been carried out. However, the domestic research on semi-permanent release agent started late, and the domestic research on disposable release agent was mainly carried out. Domestic semi-permanent release agent market has been monopolized by Daikin and other companies in Japan. With the excellent performance of semi-permanent release agent gradually emerges in recent years, domestic scholars begin to study[44].

Chem Trend developed a new generation of water based semi-permanent release agents that not only reduced the difficulty of release, but also provided a durable film. This semi-permanent, non-migrating coating was designed for all non-silicone, thermoplastic elastomers and polyaminoacetate rubber, especially for rubber processing[46].

Using perfluoropolyetheryl fluoride as raw material, Fan Yangyang[47] prepared perfluoropolyetheramide through hydrolysis and ammoniation. After emulsification, the water-based organic fluorine emulsion release agent with good chemical stability and heat stability was prepared. The product had long storage cycle and can realize semi-permanent demoulding.

Zhang Chen[48] made organic fluorine emulsion release agent with perfluoropolymer and other raw materials, conducted infrared spectrum analysis and contact angle analysis, and discussed the influence of emulsifier type, dosage and other factors on the release agent. The prepared semi-permanent release agent can rapidly cure at 25°C, and the release times can reach more than 20 times.

Li Bin [41] developed two semi-permanent release agents that could be cured at room temperature based on organosilicon and organofluorine respectively. A crosslinking silicone resin was synthesized from two kinds of silanes and ethyl orthosilicate, and silicone oil and catalyst were added to prepare silicone release agent. The effects of resin type and amount of resin, amount of silicone oil and catalyst on the properties of release agent were studied, and then the curing properties were tested, and the effects of catalyst and release agent concentration on the curing properties were studied. Perfluoropolyetheryl was hydrolyzed and aminated into perfluoropolyetheramide, and then emulsified into organic fluorine release agent. The effects of emulsifier type and dosage, additives and other factors on the properties of the release agent were studied. The two semi-permanent release agents can be cured at 25°C with good stability and high release times.

Li Zhiming [49] et al. developed a semi-permanent release agent using polysiloxane and catalyst as raw materials, which had the advantages of low curing conditions, low cost and easy-to-use, and was suitable for epoxy resin products and rubber products.

Zhu Hang [50] et al. invented a water-based semi-permanent release agent with a variety of fluorine-containing compounds as raw materials. Compared with organosilicon release agent, the surface tension is lower and the release efficiency was higher, and the stripping times of the release agent can reach 23 times.

5 Conclusion

Semi-permanent release agent not only has the characteristics of high demoulding times, no pollution, low cost, less manual cleaning. It can also ensure the smooth surface of products and molds, and its excellent performance is unmatched by disposable release agent. However, high-end products have been dependent on imports, domestic research is relatively backward, causing a huge hindrance to the production of rubber, plastic and other industries. In order to break the monopoly situation of foreign companies, save costs, improve productivity and protect the environment, it is of great significance to develop semi-permanent release agents.

Acknowledgments

Thanks to the following funds for supporting this project. Heilongjiang Provincial Natural Science Foundation of China (LH2020E089), Postdoctoral Innovation Project of Shandong Province (202103051), China Postdoctoral Science Foundation (2021M691970), University Nursing Program for Young Scholars with Creative Talents in Heilongjiang Province (UNPYSCT-2018205), Fundamental Research Fundation for Universities of Heilongjiang Province (2018-KYYWF-1598).

References

[1] Zhigang Lei. Preparation and application of fluorine-containing demoulding agent[D]. Zhejiang University of Technology,2009.

[2] Amra, Kesrin, Momin, etc. Formulation evaluation of ketoconazole microemulsion-loaded hydrogel with nigella oil as a penetration enhancer[J]. Journal of Cosmetic Dermatology, 2019, 18(6): 1742-1750.

[3] Benito J. M., Cambiella A., Lobo A., etc. Formulation, characterization and treatment of metalworking oil-in-water emulsions[J]. Clean Technologies and Environmental Policy, 2010, 12(1): 31-41.

[4] Beshkar Farshad, Khojasteh Hossein, Salavati-Niasari Masoud. Recyclable magnetic superhydrophobic straw soot sponge for highly efficient oil/water separation[J]. Journal of Colloid and Interface Science, 2017, 497(56): 57-65.

[5] Cambiella Angel, Benito Jose A., Pazos Carmen, etc. Interfacial properties of oil-in-water emulsions designed to be used as metalworking fluids[J]. Colloids and Surfaces a-Physicochemical and Engineering Aspects, 2007, 305(1-3): 112-119.

[6] Cambiella A., Benito J. M., Pazos C., etc. The effect of emulsifier concentration on the lubricating properties of oil-in-water emulsions[J]. Tribology Letters, 2006, 22(1): 53-65.

[7] Cappelletti G., Ardizzone S., Meroni D., etc. Wettability of bare and fluorinated silanes: A combined approach based on surface free energy evaluations and dipole moment calculations[J]. Journal of Colloid and Interface Science, 2013, 389(557): 284-291.

[8] Carvalho De Almeida C., De Araujo Melo D. M., Maciel M. A. M., etc. Use of methanolic extract of ixora coccinea linn for corrosion inhibitor, which comprises preparation, characterization and application of microemulsion system based on extract as corrosion inhibitor, containing oil phase and aqueous phase. Maciel M a M: 英国.

[9] Cerpnjak Katja, Zvonar Alenka, Vrecer Franc, etc. Development of a solid self-microemulsifying drug delivery system (smedds) for solubility enhancement of naproxen[J]. Drug Development and Industrial Pharmacy, 2015, 41(9): 1548-1557.

[10] Chaydarreh Kamran Choupani, Shalbafan Ali, Welling Johannes. Effect of ingredient ratios of rigid polyurethane foam on foam core panels properties[J]. Journal of Applied Polymer Science, 2017, 134(17): 224-675.

[11] Chouhan P., Saini T. R. D-optimal design and development of microemulsion based transungual drug delivery formulation of ciclopirox olamine for treatment of onychomycosis[J]. Indian Journal of Pharmaceutical Sciences, 2016, 78(4): 498-511.

[12] deDamborenea J., Bastidas J. M., Vazquez A. J. Adsorption and inhibitive properties of four primary aliphatic amines on mild steel in 2m hydrochloric acid[J]. Electrochimica Acta, 1997, 42(3): 455-459.

[13] Erdal M. Sedef, Ozhan Gul, Mat M. Cem, etc. Colloidal nanocarriers for the enhanced cutaneous delivery of naftifine: Characterization studies and in vitro and in vivo evaluations[J]. International Journal of Nanomedicine, 2016, 11(6): 1027-1037.

[14] Figueiredo Leandro, Bandeira Pedro, Mendes Adelio, etc. Use of fluoropolymer permanent release coatings for molded polyurethane foam production[J]. Journal of Coatings Technology and Research, 2012, 9(6): 757-764.

[15] Fu Yanhui, Li Zhijun, Liu Qinqin, etc. Construction of carbon nitride and mos2 quantum dot 2d/0d hybrid photocatalyst: Direct z-scheme mechanism for improved photocatalytic activity[J]. Chinese Journal of Catalysis, 2017, 38(12): 2160-2170.

[16] Sirtl Erhard , . Über die reduktion von chlorsilanen mit wasserstoff[J]. Reuschel, 2004, 332(3-4): 113-123.

[17] Fu Yanhui, Liang Wei, Guo Jinqiu, etc. Mos2 quantum dots decorated g-c3n4/ag heterostructures for enhanced visible light photocatalytic activity[J]. Applied Surface Science, 2018, 430(3): 234-242.

[18] Kabakci Elif, Sayer Goksin, Suvaci Ender, etc. Processing-structure-property relationship in rigid polyurethane foams[J]. Journal of Applied Polymer Science, 2017, 134(21): 354-554.

[19] Kim S. H., Park H. C., Jeong H. M., etc. Glass fiber reinforced rigid polyurethane foams[J]. Journal of Materials Science, 2010, 45(10): 2675-2680.

[20] Kumar Neeraj Shishu. D-optimal experimental approach for designing topical microemulsion of itraconazole: Characterization and evaluation of antifungal efficacy against a standardized tinea pedis infection model in wistar rats[J]. European Journal of Pharmaceutical Sciences, 2015, 67(7): 97-112.

[21] Li Caifu, Mei Zhen, Liu Qian, etc. Formation and properties of paraffin wax submicron emulsions prepared by the emulsion inversion point method[J]. Colloids and Surfaces a-Physicochemical and Engineering Aspects, 2010, 356(1-3): 71-77.

[22] Li Guo, Dong Lei, Bai Zhu'an, etc. Predicting carbonation depth for concrete with organic film coatings combined with ageing effects[J]. Construction and Building Materials, 2017, 142(6): 59-65.

[23] Kim Kwang J., Shahinpoor Mohsen A novel method of manufacturing three-dimensional ionic polymer–metal composites (ipmcs) biomimetic sensors, actuators and artificial musclesj[J]. Polymer, 2015, 43(3): 797-802.

[24] Masuda Jun, Yan Jiwang, Zhou Tianfeng, etc. Thermally induced atomic diffusion at the interface between release agent coating and mould substrate in a glass moulding press[J]. Journal of Physics D-Applied Physics, 2011, 44(21): 345-677.

[25] Haowei Zhao. Preparation of self-emulsifying block silicone oil[J]. New Chemical Materials, 2019, 2(1): 1-4.

[26] Otsuka Yuma, Hiwasa Shin, Taniguchi Jun. Development of release agent-free replica mould material for ultraviolet nanoimprinting[J]. Microelectronic Engineering, 2014, 123(53): 192-196.

[27] Peres Francois, Mofakhami Arthur Rapid tooling from protoypes: The cool-spray mold-making method[J]. Proceedings of SPIE - The International Society for Optical Engineering, 1998, 3517(11): 79-90.

[28] Scaffaro R., Maio A., Gulino F. E., etc. Bilayer biodegradable films prepared by co-extrusion film blowing: Mechanical performance, release kinetics of an antimicrobial agent and hydrolytic degradation[J]. Composites Part a-Applied Science and Manufacturing, 2020, 132(76): 453-658.

[29] Serre C., Vayer M., Erre R. Behavior of internal mold release agent during bmc thermosets composites cure and aging[J]. Journal of Materials Science Letters, 2001, 20(21): 1989-1991.

[30] Lotierzo A., Pifferi V., Ardizzone S., etc. Insight into the role of amines in metal working fluids[J]. Corrosion Science, 2016, 110(3): 192-199.

[31] Migahed M. A., Abd-El-Raouf M., Al-Sabagh A. M., etc. Corrosion inhibition of carbon steel in acid chloride solution using ethoxylated fatty alkyl amine surfactants[J]. Journal of Applied Electrochemistry, 2006, 36(4): 395-402.

[32] Miyata Takahiro, Matsumoto Kozo, Endo Takeshi, etc. Synthesis and radical polymerization of styrene-based monomer having a five-membered cyclic carbonate structure[J]. Journal of Polymer Science Part a-Polymer Chemistry, 2012, 50(15): 3046-3051.

[33] Negi Ripudaman Singh, Bisht Rajendra Singh, Singh Raj K. Physico-mechanical and antibacterial properties of pine gum/epoxy composites with/without silver nanoparticles[J]. Macromolecular Materials and Engineering, 2019, 304(7): 335-577.

[34] Neis Franciele Antonia, de Costa Fernanda, Fuller Thanise Nogueira, etc. Biomass yield of resin in adult pinus elliottii engelm. Trees is differentially regulated by environmental factors and biochemical effectors[J]. Industrial Crops and Products, 2018, 118(3): 20-25.

[35] Padula C., Telo I., Di Ianni A., etc. Microemulsion containing triamcinolone acetonide for buccal administration[J]. European Journal of Pharmaceutical Sciences, 2018, 115(2): 233-239.

[36] Pal Kaushik, Rajasekar R., Kang Dong Jin, etc. Effect of fillers on natural rubber/high styrene rubber blends with nano silica: Morphology and wear[J]. Materials & Design, 2010, 31(2): 677-686.

[37] Guanghui Yuan. Study on preparation process of organic silicon release agent[D]. Qingdao University of Science and Technology,2013.

[38] Mingming Xu. Synthesis and application of functional organosilicon compounds[D]. Tianjin Polytechnic University,2017.

[39] Chengyong Jiang. Silicone release agent and its application[J]. Plastics Industry, 2008, (02): 74-75.

[40] Runlin Li. Development and characteristics of water - based semi-permanent release agent[J].

Rubber Reference, 1992, (05): 22-28.

[41] Bin Li. Study of room temperature curing semi-permanent release agent[D]. Dalian Polytechnic University,2015.

[42] Hongsheng Guo. Development and characteristics of water - based semi-permanent release agent[J]. Rubber translations, 1993, (03): 19-24+3.

[43] Liying Wu, jianjun Gao, Wugang Jin. The release agent of composites metalized by transfer[J]. Colloids and Polymers, 2001, (04): 21-22.

[44] Bin Li, Hongyi Hai, Jianlou Ge, etc. Preparation and performance of semi-permanent release agent cured at room temperature[J]. Journal of Dalian Polytechnic University, 2015, 34(05): 345-348.

[45] Haibin Zhou. Use semi-permanent release agent to help release mold and prevent mold contamination[J]. World rubber Industry, 2007, (01): 33-36.

[46] Xuezhong Tu. Semi-permanent release agent for precision model products[J]. Rubber Industry, 1996, (03): 184.

[47] Yangyang Fan, Bin Lin, Chen Zhang, etc. Preparation property of semi permanent mold release agent at room temperature[J]. New chemical materials, 2018, 46(03): 200-202+206.

[48] Chen Zhang, Bin Li, Yanjun Liu. Preparation of semi-permanent release agent cured at room temperature[J]. Technology Outlook, 2017, (4): 38.

[49] Zhiming Li, Wentao Xing, Jianyan Xu, etc. Preparation method and application of a release agent. 2010.

[50] Hang Zhu, Xinliang Peng. A water-based semi-permanent organic fluorine release agent. 2019.

MATEC Web of Conferences **353**, 01005 (2021) https://doi.org/10.1051/matecconf/202135301005
AMME 2021

Six Sigma application for quality improvement of the hot press in process

Xin Li *

Institute of Management, Tianjin University of Technology, Tianjin, China

Abstract. Six Sigma is a systematic improvement method, a management approach designed to continuously improve corporate business processes and achieve customer requirement. This paper presents a case study conducted at a cell phone manufacturing company that intended to use the DMAIC method to carry out a structured approach to Six Sigma projects with fewer defective units generated by the cell phone hot press in process. In this process, Boss cracking and torque force failure led to high scrap costs and line downtime, which ultimately affected the supply of the assembly line. To this end, the Six Sigma approach was applied starting with the problem being defined, measured, and exhaustively analyzed to determine the root cause: poor injection molding structure, metal insert not perpendicular to the hole, improperly set molding machine parameters, and core/cavity dimension variation. A series of improvement measures addressing these factors increased the quality level of the process by improving the mold design to increase venting, optimizing the product structure to add the guide chamfer, and developing a standardized molding parameter setup procedure. Methods to maintain process control were also identified and implemented. The use of several quality tools and the use of Six Sigma methods also made the process more consistent.

1 Introduction

Due to constant changes in customer needs, new markets, innovations and other external factors, organizations are under constant pressure to improve the quality of their existing processes and to develop new processes to meet market trends. For this purpose, powerful methods of quality management, such as Six Sigma [1][2][3], can be used. According to Linderman et al [4], Six Sigma can be seen as a systematic, organized, and customer-oriented approach that aims to improve the performance and quality of processes, products, and services by using statistical techniques and scientific methods to analyze data and make decisions. The application of Six Sigma to repetitive processes has brought significant benefits to companies. This paper describes a project undertaken by a cell phone manufacturing company whose goal was to use Six Sigma methods to reduce the number of defective cells in a process that is a common part of cell phone manufacturing: the hot press in process. It has been found that in the hot press in process, if the process parameters are not set properly there are cosmetic defects such as failure of torque force or even cracking and higher scrap costs and reduced productivity, which affects the supply of the final assembly line. Through this project, we intend to reduce the cost of quality in this process and increase the number of products that are produced in the first time, thus not affecting the supply of the final assembly line. In this paper, we show the significant advantages obtained in the hot press in process by applying the DMAIC [5]approach and quality tools of the Six Sigma methodology. The paper is organized as follows. Section 2 describes the hot press in process. In section 3, the project is reported according to the five steps of the DMAIC method. The final section presents the conclusions of the implementation of the methodology in the hot press in process.

2 Hot Press in Process

The cell phone manufacturing process is divided into ten stages according to the finished form of material processing. The first stage to fifth stage is mechanical parts manufacturing and simple assemblyng plastic parts; the sixth stage is the manufacturing of the motherboard and other small cards ; the tenth stage is the system integration and final test including installation of software packages, such as operating system software installation, operating system testing, and finally packaging and labeling. The hot press in process is in the first to fifth stage which is based on the high temperature heating state of the molded Boss feature for secondary heating, high temperature to change the physical properties of plastic, then use external force to press the brass insert with thread, natural cooling resin wrapped thread to form a stable structure. This series of processes is completed by a special hot machine. The Boss feature with press in brass insert is used for the assembly and fixing of cell phone body and cover, as the excellent physical properties of the brass inserts provide better mechanical strength and repeated use.

* Corresponding author: lynlee@stud.tjut.edu.cn

3 Six Sigma Project

The Six Sigma project is presented next following the DMAIC steps.

3.1 Define

In the first phase, the project team was identified. The team consisted of a team leader, a black belt from the engineering department, process owner, process factory specialist, process maintenance specialist, two quality leaders and a sponsor. After the project team was selected, the following steps were performed.

Formalize the Six Sigma project by defining the name of the project, the time period of the data to be analyzed and the schedule of the work to be performed. The expectations of the project were exposed, especially in terms of the benefits to the company. A project charter containing this information is published.

Project Schedule Plan. The project schedule is particularly important in a quality improvement project. The ability to carry out the various elements of work effectively and to meet expectations is critical to the success of this project. In order to clearly understand the start and end time of each task and to better control and improve the project schedule, this paper uses Gantt charts[6] to plot the schedule control milestone nodes and the start and completion time of each phase of the task

Process characteristics. The SIPOC[7] diagram is used to identify the inputs that may affect the hot press in process and the outputs that need improvement. This tool allows focusing on the factors that lead to process changes.

Definition of project indicators. Defects were considered to be torque force that exceeded the upper and lower specification limits and poor appearance of cracked Boss columns. A number of other metrics were also used, such as the number of phone cases and the associated loss of scrap costs

Calculation of Sigma levels. The initial sigma level was calculated based on data from a 6-month time interval. The initial value was 2.53. In order for the project to be considered successful, the goal to be achieved was to reduce the DPMO (defects per million opportunities) by at least 80%, which corresponds to a sigma level of 3.38.

3.2 Measurement

Correct data is the result to be obtained from the measurement phase of the work process. The correctness of the data depends on the reliability of the quality of the measurement process, i.e. the quality of the measurement system. Therefore, before analyzing the data, the reliability of the data source should be ensured, i.e., the reliability of the current exterior inspectors' judgment of the Boss feature cracking problem and the accuracy of the measurement of the torque force by the measuring personnel should be assessed. Because the data for these two indicators are attribute data and variable data, we used Assessment Agreement and Gage R&R[8] analysis for these two parts of the measurement system, respectively. For the Boss feature cracking problem, we mixed the cracked phone top cover and the intact phone top cover with a good number and did visual identification by three different final inspectors, who were screened and specially trained. Each of them did 2 tests in order to verify the repeatability and reproducibility of the test. Samples and data were also collected for the torque force. Then, we did Assessment Agreement (Figure 1). and Gage R&R analysis (Figure 2). on the collected attribute data and variable data, respectively, with the help of Minitab software. The experimental results prove that the current measurement system is acceptable.

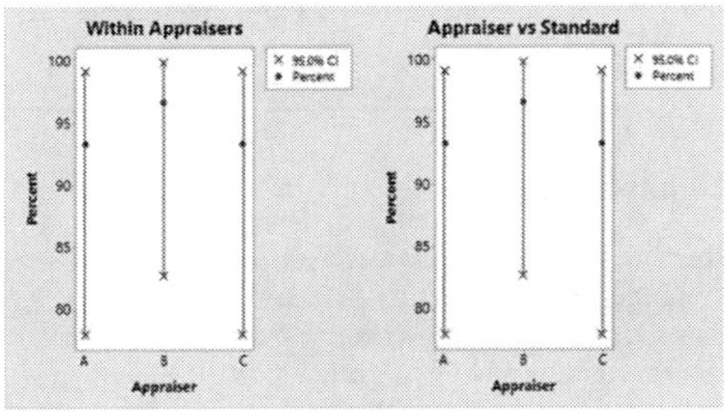

Figure 1. Schematic diagram of Assessment Agreement

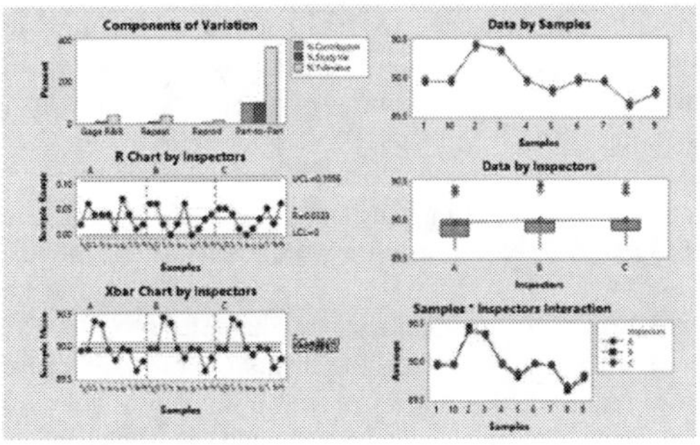

Figure 2. Schematic diagram of Gage R&R analysis

3.3 Analysis

The analysis phase is mainly to determine the key factors affecting the process output through data analysis, i.e., to identify the key influencing factors of the process. In this paper, we mainly use cause-effect matrix diagram, failure mode analysis[9] and double ratio hypothesis testing. In order to understand the real causes of cracking and torque force failure in the process of hot press in of cell phones, team members used brainstorming method to start group discussion based on the process flow chart, and summarized the possible influencing factors from people, machine, material, method, environment and other impossible aspects after analysis, which appeared in the form of cause-effect matrix diagram. The input variables that may affect insert installation and hot press in were classified from the cause-effect matrix, and in order to further find the key factors caused by the defect, the input variables with high scores in the process of hot press in: temperature of the hot press in machine, material size, location of the hot press in fixture, strength of the material knit line and the method of installing the brass insert were

https://doi.org/10.1051/matecconf/202135301005

analyzed with the FMEA to arrive at the RPN greater than 100 in Table 1. four critical factors. The team further confirmed that the four critical factors had an effect on Boss feature cracking and torque force failure by collecting data in small batches using a double ratio hypothesis test.

Table 1. Critical fators

Numbers	Factors description
X1	Poor injection molding structure
X2	Metal insert not perpendicular to the hole
X3	Improperly set molding machine parameters
X4	Core/cavity dimension variation

3.4 Improvement

Poor injection molding structure

Combined with the actual defective product occurrence status, the defective location of the model check, by comparing the cracking defective products within a week analysis found that the insert cracking mostly concentrated in one area. From the observation of the occurrence of the form, it is suspected that the location of the knit line, the need for mold flow analysis and short shot test mold analysis. Through the observation after mold flow analysis, the cracked area coincides with the location of the bond line we suspect, which will largely affect the strength and appearance of the product. For the weak structural strength at the bond line, we use cut through into the sub to improve the strength of the insert hole.

Metal insert not perpendicular to the hole

Before the improvement, the installation of copper nails only required to put the insert into the screw hole, after the improvement, we required to put the insert into the screw hole and press the insert downward to confirm the insert into place, and then carry out the hot press in operation at the next work station. And set the insert installation station as a key station, fixed staffing, by skilled operators to carry out special training and qualified to work. In addition, the manufacturing engineer will take the lead in designing and developing a positioning fixture with guiding function, so that the insert can be installed vertically even if the operator does not operate properly occasionally. Finally, from the product design point of view, we hope to add a chamfer with guiding function to the screw hole, and this plan needs to be sent back to the customer for confirmation before implementation.

Improperly set molding machine parameters

Through the analysis of the product of torque failure, as well as observation of the cracking of the cell phone that occurred many times during the assembly production process, we found that the main reason was the high temperature of the hot runner during the molding production and the non-standard operation during the production. After review the datasheet from raw material supplier, the manufacturing engineers set reasonable molding conditions: the molding machine material temperature should not be higher than 275 degrees Celsius. If you need to adjust the material temperature (more than 275 degrees Celsius), you must obtain the consent of the quality control worker before adjustment. When the machine is stopped for more than 3 minutes, the material must be cleared and the residual material in the hot runner must be eliminated.

Core/cavity dimension variation

In the process of mass production, the mold is used for a long time, the internal structure and parts wear, the size of the product in the corresponding position after the use of mold molding also has a corresponding variation. Reflected in this project is the variation of the size of the hole at the location of the hot insert. According to past experience, the dimensions guaranteed by the mold are often not included in the SIP process inspection specification, which eventually led to the problematic products output to the assembly line, resulting in torque force cracking after the insert was hot press in. The quality engineer designed and made the Go no Go inspection jig according to the customer specifications.

3.5 Control

Standardize the improved process settings

Develop standardized inspection specifications, add them to SIP for standardized control according to the improvement content set in the improvement phase, and issue the revised standard process inspection specifications to each inspector.

Use the statistical process control to monitor the factor level settings or the indicators of project

Process has been in a state of fluctuation, collect data from one month after implemented the improvement actions and draw the Figure 3 which is a P chart from SPC tools, all data are fluctuate in the mean line. The defects rate reduced to 97% and sigma level successful promote to 3.38.

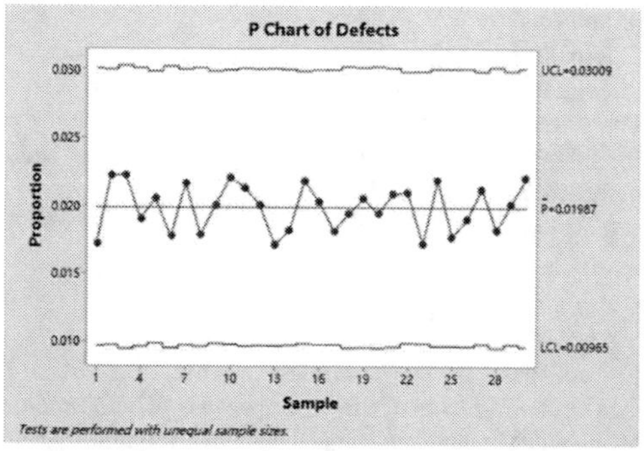

Figure 3. Schematic diagram of P Control Chart

4. Conclusion

This paper describes the completion of a Six Sigma project that consisted of reducing the number of defective units generated during the hot press in insert process for cell phones by applying Six Sigma methods to reduce the inherent cost of quality without affecting the supply of the final assembly line.

References

1. J. Antony R. Banuelas. Design for Six Sigma: A Breadthrough Improvement Business Strategy for Achieving Competitive Advantage[J]. Manufacturing Engineering.2002, 81(1):24-26.
2. F. Breyfogle, Implementing six sigma: smarter solutions using statistical methods. John Wiley & Sons, 2003.
3. M. Kumar, J. Antony, B. Cho, Project selection and its impact on the successful deployment of Six Sigma, Business Process Management Journal, 15(2009) 669–686.
4. Linderman K, Schroeder R G, Zaheer S, et al. Six Sigma: a goal-theoretic perspective[J]. Quality Control & Applied Statistics, 2003, 21(2):193-203
5. Lenka Girmanová et al. Application of Six Sigma Using DMAIC Methodology in the Process of Product Quality Control in Metallurgical Operation[J]. Acta Technologica Agriculturae, 2017, 20(4) : 104-109.
6. Masahiro Shibuya. Production Planning and Management Using Gantt Charts[J]. Journal of Mechanics Engineering and Automation,2021,11(3):
7. Brown Charlie. Why and how to employ the SIPOC model.[J]. Journal of business continuity & emergency planning, 2019, 12(3) : 198-210.
8. Jay Arthur. Gage R&R[J]. Quality, 2017, 56(13) : 20-21.
9. Lorenzi Clarice Inês and Ferreira Joao Carlos Espindola. Failure mapping using FMEA and A3 in engineering to order product development[J]. The International Journal of Quality & Reliability Management, 2018, 35(7) : 1399-1422.

Exploration on Application of High Pressure Water Jet Cleaning Technology

Xinjian Xu [1,*], Zhaoxing Meng [2] and Haiqiang Lv [1]

[1] Shandong Xiehe University, Shandong, China
[2] Jinan Vocational College, Shandong, China

Abstract: at present, the three-dimensional cleaning machine involved in tank car cleaning technology in China has full-automatic function, which includes two main structures: three-dimensional positioning machine and high-pressure cleaning machine. However, the tank car cleaning process used in the railway field still has many shortcomings. For example, serious pollution, serious energy consumption, insecurity, etc. In order to improve such defects, it is necessary to introduce new processes and improve the previous cleaning processes. The introduction of new technology can reduce environmental pollution and save capital cost. It can adapt to the development of the times and meet people's needs. Therefore, this paper expounds and analyzes the high-pressure water jet cleaning technology in railway tank cars from the aspects of application and improvement, and puts forward the corresponding improvement measures for reference.

1. Introduction

Through the installation of the nozzle, the target distance between the nozzle and the inner wall of the tank car can be kept appropriate, and the automatic and efficient cleaning of high-pressure water jet is realized. This kind of new technology and process has been rising rapidly. The core is high-pressure water jet technology. This automatic control system integrates machine, pump and liquid. It has certain advantages in structure, operation and automation, and has reliable and complete product quality, which can make the effect safer, energy-saving and environmental protection.

2. Overview and technical advantages

2.1 Basic overview

In view of the rapid rise of China's petrochemical industry, petrochemical products tend to be cleaner and large-scale transportation. Railway tank car is its main means of transportation. However, due to the flammable and explosive characteristics of products in the petrochemical field, there are certain requirements for safety factor, pollution degree and cost in the transportation stage.In order to meet the needs of customers and make the tank car have the advantages of cleaning and environmental protection, the tank washing station needs to be designed to clean the internal residues in the process of transportation of petrochemical industry.

At present, there are two operation modes of tank washing process in China. One method is manual cleaning process. That is, traditional steam fumigation, low-pressure fresh water washing and hot air drying. The other method is mechanical cleaning method. Because the former has the disadvantages of small scale, high energy consumption, low efficiency, serious environmental pollution and great personal harm to operators, many large enterprises and petroleum product transfer stations adopt mechanical cleaning method.

2.2 At present, the main high-pressure water jet cleaning equipment adopts the high-pressure water jet technology during the cleaning of iron tank truck, which is mainly composed of the following core equipment:

(1) High pressure water supply equipment.It mainly involves water tank, high-pressure water pump, water valve and other equipment.In the process of operation, the industrial water passing through the filter screen will enter the water tank, and the clean water will be brought into the high-pressure water pump by the manual valve or solenoid valve. After the internal water reaches the pressure value of 20 ~ 70MPa under the action of the high-pressure water pump, it can be sprayed to the pipe wall by the high-pressure water pipe and washer, and the internal residues will be hitAfter the action of high-pressure water such as erosion and cutting, the equipment can be cleaned.
(2) Clean the equipment.It mainly involves three-dimensional crane, cleaning machine and other equipment.These are the core high-pressure water jet cleaning equipment. Dozens of manufacturers at home and abroad have studied and produced such equipment and widely used it in cleaning.

* Corresponding author: 302610821@qq.com

3. Technical advantages and disadvantages

3.1 Advantage

High pressure water jet technology, a cleaning process, originated in the 1970s. It forms the jet through the injection of specific pressure water through the nozzle, can impact and cut the pollutants and rust left in the railway tank car, and can clean the substances left in the tank car.Therefore, the cost saving and environmental protection effect are greatly improved.Compared with the previous cleaning process, tank car cleaning has the following specific advantages for the use of high-pressure water jet:

(1) Elimination of air pollution.Because it does not need steam operation, there will be no materials left in the tank car volatilized with steam in the air, and the materials left in the tank car treated by the sewage plant are washed down by high-pressure water. Therefore, it can provide better working environment for workers and keep the atmosphere healthy.

(2) Efficient washing and irrigation.For the use of high-pressure water jet technology during washing and irrigation, generally speaking, it takes 40min to clean a vehicle and 60 ~ 70min / set to clean a slightly more difficult tank car.The use of high-pressure water jet technology in the washing and irrigation stage can improve the car washing capacity of each washing station by 4 ~ 5 times.

(3) Significantly reduced habitual energy consumption and cost.For the use of high-pressure water jet technology in the washing and irrigation stage, the water and power consumption of each unit are 4 ~ 6T and 140kW • h respectively. Compared with steam tank washing, it achieves more than 90% energy-saving effect, and each unit is reduced by 700 ~ 800 yuan.

3.2 Defect

(1) The height is not matched.For the production of 11m high and 2m wide three-dimensional crane in a unit, if the washing and irrigation station wants to transform the original steam car washing technology through the installation of high-pressure water jet cleaning equipment during washing and irrigation, it can not be realized.This is because this kind of washing and irrigation station has a plant 7 ~ 9m high and a platform more than 1m wide.

(2) Uneven cleaning quality makes it impossible to clean point by point.If the cleaning machine produced by a unit can control the three-dimensional cleaning head of the arc motion track, the three-dimensional cleaning head in the tank car will spray water on the arc, and all points inside the tank car will be subject to different flushing intensities.

(3) Serious waste of high-pressure water.The use of four point cleaning by a unit, that is, the three-dimensional cleaning head in the tank car cleans at each point inside. The dwell time at each point is a few minutes. There is also the problem of uneven cleaning. Only part of the high-pressure water can be effectively used, and the other part is ineffective.

4.Transformation scheme

The high-pressure cleaning machine relates to the main components such as travel switch, frame, counterweight and rocker arm.The elbow end connected to the three-dimensional cleaning nozzle is C-shaped. Through the action of the feed reducer on the horizontal rocker arm, the C-shaped elbow can be sent into the tank car. Through the simultaneous control of its travel through the switch and program, the three-dimensional cleaning nozzle in the tank car can be cleaned at four points.After unilateral cleaning of the tank car, it is necessary to rotate the horizontal feeder 180 ° , and then clean the other side of the tank car.Through the automatic extraction of oily sewage by relevant institutions, the pumping sewage can be automatically transmitted to the bottom of the tank car through stainless steel pipe. The external pipe for pumping oily sewage can extract the oily sewage and sewage after installing the joint.The high-pressure cleaning machine involves explosion-proof equipment such as instrument junction box and distribution box.

5. Transformation design

5.1 Design content

In the process of cleaning, the c-bend can rotate 90 ° under the action of motor and a series of deceleration measures, and then the c-bend swings under the action of reducer.In the process of sending the three-dimensional cleaning nozzle, it is necessary to stop at four points for cleaning.During cleaning, the number of cleaning points to stay and the cleaning time of each point shall be set based on the actual demand;If the motor and a series of deceleration measures are used to reverse 180 ° rotation of the C-shaped elbow after cleaning one end of the tank car, it is necessary to ensure that the three-dimensional cleaning nozzle has returned to the tank mouth on the C-shaped elbow, and then clean the other end of the tank car.After cleaning the tank car, for the 90 ° reverse rotation of the C-shaped elbow, it is necessary to ensure that the three-dimensional cleaning nozzle has returned to the tank mouth on the C-shaped elbow, raise the three-dimensional positioning machine to the original position, and prepare for cleaning the next tank car.The water pressure in the cleaning system can reach 34.5mpa and theoretically reach the flow of 110L / min.The eccentric setting of the three-dimensional cleaning nozzle to the two nozzles is mainly used to spray high-pressure water. After the force couple is formed, it can rotate the nozzle. The gear transmission fixed on the nozzle shaft can rotate the shell of the three-dimensional cleaning nozzle.Therefore, the water column can be sprayed in all directions in the tank.By feeding the three-dimensional cleaning nozzle in the horizontal direction, the movement of the three-dimensional cleaning nozzle in the tank car can always be kept near the center line. Through accurate positioning, the clean water can be saved, the energy can be used efficiently, the cleaning speed and cleanliness can be improved, and the cleaning of the tank car that is difficult to clean can be realized.

5.2 Working principle and process

High pressure jet and robot are the core technologies of this technology.Its subsystems include high-pressure cleaning system and automatic control system.Its core system is the high-pressure cleaning system, which involves flood device, high-pressure pump and high-pressure cleaning machine.In the process of water supply, the water supply device can reach a pressure of 20 ~ 30MPa through the pressurization of the high-pressure pump. Under the high pressure, the cleaning machine can spray water into the inner wall of the tank car together with the three-dimensional washer. The internal residues can clean the equipment after being hit, eroded, cut and other high-pressure water.

5.3 Process description

After the tank car that has not been cleaned reaches the designated position under the traction of the locomotive, if there are many residues in the tank car in winter and the residues have serious icing, it is necessary to heat them with steam first, and then extract the residues; Finally, it can be cleaned.

After cleaning, waste water can be pumped into the sewage tank through the use of vacuum pump. When the water level in the tank is equal to the preset liquid level, or after cleaning a tank truck, it is necessary to automatically open and close the solenoid valves of the compressed air pipeline and vacuum pipeline respectively, and turn off the vacuum pump, and then the sewage pressed into the tank is treated by the sewage treatment system.

After the cleaning of the first and second tank cars is completed, the tractor shall pull them to the next two positions for cleaning. After the tank car is cleaned and the tank cars at the upper two stations are cleaned, it can be towed to the next two stations for drying, and the tank cars cleaned at the upper two stations can be towed to this station for cleaning. The dried tank car is towed to the next two platforms for supplementary drying and inspection.

After applying this technology to the tank washing stations of the three companies, the car washing efficiency is greatly improved. For general original light oil tank cars, the actual cleaning time of a single car is shortened to 12 ~ 15min, and the maximum daily cleaning capacity of one bench is 15 vehicles; At the same time, the energy consumption, environmental pollution and personal harm to workers are greatly reduced, and the economic and social benefits are very obvious.

6. Suggestions and application prospects

In terms of economic and social benefits, this technology is a practical new technology worthy of vigorous promotion. Compared with the latest technical progress at home and abroad, the system can also be improved in three aspects: first, increase the advanced treatment facilities of tank washing sewage, reuse the treated purified water and save water; Second, metal mechanical drying facilities are adopted to avoid the fiber falling into the tank car during the drying process of the canvas air duct, resulting in secondary pollution, and improve the drying efficiency of the tank car; Third, adopt mature, reliable and economical oil and gas recovery technology to further reduce air pollution.

In China, there are tank car cleaning operations in railway, petroleum, petrochemical and chemical enterprises. Most of the existing tank washing stations in China still use the method of manual car washing. Moreover, in the railway system, due to the maintenance of tank cars and the external cleaning of tank cars, energy waste and environmental pollution are very serious. If the existing tank washing stations are reconstructed, the economic and social benefits will be considerable.

7. Conclusion

In order to improve the defects of tank car cleaning process, it is necessary to introduce the new process and improve the previous cleaning process. The introduction of new technology can reduce environmental pollution and save capital cost. It can adapt to the development of the times and meet people's needs. Therefore, this paper puts forward the improvement scheme of high-pressure water jet cleaning technology for reference.

Acknowledgments

This work was financially supported by Scientific Research Projects of Shandong Higher Education Institutions (J18KB043).

References

1. State Administration of quality and technical supervision. Tsgr0005-2011 safety and Technical Supervision Regulations for mobile pressure vessels [S]. Beijing: Xinhua press, 2011
2. Zhang Dongliang, Chao daiqing. Measures and effects of energy conservation and emission reduction in tank car cleaning [J]. Equipment and corrosion protection, 2010,38 (4): 329-332
3. Li Jie, Li Duomin, Hou lailing. Current situation and development trend of industrial cleaning in China [J]. Guangdong chemical industry, 2008 (12): 69-70
4. Liu tingcheng, fan Xiaohong. Application of high pressure water jet cleaning equipment [J]. China Equipment Engineering, 2013, 02:8-10
5. Zhou Yawen, Xu Baocai, Han Fu. Application status and development trend of industrial cleaning technology in China [J]. China washing products industry, 2010, 01:33-36
6. Cheng Zhixue, Jiang Dianhong, sun Yongjun. Railway tank car cleaning robot [P]. Chinese patent: ZL 200610092305.5, 2007-2-28
7. Shen Wendong, Song Fei. Application of high pressure water jet cleaning technology for railway chemical tank cars [J]. Chemical design, 2004, 14 (2), 26-28.
8. He Mingchao. Selection and application of railway tank car cleaning technology [J]. Engineering technology, 2015 (8), 17.

9. Li Shu. Application of high pressure water jet technology in cleaning railway tank cars [J]. Cleaning world, 2004 (5) , 16-20. [4] Xu Xiaoxia, Xu Bin. Effects of occupational exposure to amino and nitro compounds of benzene on human health [J]. Occupation and health, 2000 (3), 12-13
10. Shen huanghong, Zhang Gang, Zhang Yifu, etc Overview of potassium nitrate production unit with an annual output of 20 kt Progress in chemical industry, 2002 (7): 506 ~ 508
11. Compiled by the Editorial Committee of chemical engineering manual Handbook of Chemical Engineering (2) Beijing: Chemical Engineering Publishing house, Beijing: 1989
12. P. L. Kaye . Investigation of Erosion Processes as Cleaning Mechanisms in the Removal of Thin Deposited Soils[J]. America: Wear,1995, 186(8) : 413~ 420
13. M. C. Leu. Mathemtical Modeling and Experimental Verification of Stationary Water Jet Cleaning Process[J] . America: Journal ofManu facturing Science and Engineering, 1998, 120(8) : 571~ 579
14. Shen Zhonghou Water jet theory and technology [M] Dongying: University of Petroleum Press, 1999
15. Zhang Lanfang, Fei Jianguo Mechanism and parameters of dirt cleaning by high pressure water jet [J] Chemical cleaning, 1997,13(5):6~9
16. Chen Yufan, Xu Xiaodong On the improvement of cleaning efficiency of high pressure water jet [J] Beijing: Chinese Journal of safety science, 1999, 9 (Supplement): 36 ~ 40

MATEC Web of Conferences **353**, 01007 (2021) https://doi.org/10.1051/matecconf/202135301007
AMME 2021

Correspondence analysis method of anti-slide pile information monitoring and slice model

Wanqi Wang*

China Academy of Railway Sciences Corporation Limited Institute of Computing Technology, Beijing, 100081 China

Abstract: In order to effectively grasp the real-time working status of anti-slip piles and the reinforcement effect of slopes, timely information monitoring of anti-slip piles is carried out, and the reinforcement range is modeled, and the analysis and evaluation method of slicing model numerical calculation and information monitoring data is proposed, and the information monitoring data is compared with numerical simulation to achieve effective grasp of By comparing the information monitoring data with the numerical simulation, we can effectively grasp the timely working condition of the anti-slip pile and the reinforcement effect of the soil around the pile. This is a very important engineering practice for the evaluation of anti-slip pile reinforced slope areas.

Keywords: anti-slip piles; information monitoring; numerical model; evaluation

1 Introduction

Lei Wenjie et al [1] conducted numerical simulation of slope reinforced by anti-slip piles based on the finite element strength reduction method, and explored the influence of pile position on the stability safety coefficient, location and shape of the most dangerous sliding surface of the slope The relationship between the pile spacing and the friction angle within the soil as well as the pile spacing and the slope inclination was analyzed by Zhao Minghua et al [2] based on the sloping arch effect of the soil between piles; Nian Yankai et al [3,4] used the upper and lower limit theorems of plastic limit analysis to analyze the anchorage depth and the ultimate earth pressure on the pile side of the anti-slip pile considering the influence of various parameters in the slope; Ausilio et al [5] obtained the relationship between the optimal position of anti-slip piles and the slope safety factor by analyzing the stability of slopes with and without piles, and determined the critical sliding surface and safety factor (Eq.) for slopes without piles; Hassiotis et al [6] performed stability analysis of slopes based on the friction circle method, and introduced the stability number considering anti-slip piles to determine the new most dangerous sliding surface and safety factor; Won [7] used a shell unit simulated the interaction relationship between the anti-slip pile and the soil body, obtained the safety coefficient of the slope under different pile stiffness and pile head restraint conditions, and analyzed the shear force on the contact surface, and proposed the location of the point corresponding to the slip crack surface at the pile setting at the maximum shear force of the pile body; Dai Zihang [8] studied and deduced the different distribution of the landslide thrust and soil and rock resistance based on the model test of anti-slip pile and field test results in China functions; Song Yingjie et al [9] reviewed the theories of domestic and foreign methods on stability analysis of anti-slip column reinforced slopes and compared and analyzed their advantages and disadvantages; Liu Bo et al [10] used FLAC~(3D) for numerical simulation to study the mechanism of anti-slip pile overtopping failure of a high slope under rainfall conditions and made trial calculations on the reasonable location of anti-slip piles; Liu Tao et al [11] constructed a three-dimensional multilayered landslide body model, analyzed the formation process of the soil arch behind the pile and the four stages of soil arch evolution, and proposed a strip division method applicable to multilayered landslides.Zhang Youjun et al.[12] analyzed the various characteristics of landslides during the sliding, sliding acceleration and sliding mitigation stages through the observation data of landslide surface and support structure (anti-sliding pile) displacement, anti-sliding pile stress monitoring, landslide body deep displacement monitoring, groundwater level monitoring and fracture inspection. The monitoring methods complement each other and provide a comprehensive and accurate understanding of the state of the landslide body, which provides a reliable basis for the information design and safe construction of landslide management. By monitoring the underground displacement during the construction of the landslide, Long Wanxue et al.[13] used hyperbolic exponential smoothing, BP neural network and polynomial regression for tracking prediction, and Verhulst inverse function model for time prediction, analyzed the sliding speed and deformation stage of the landslide, combined with rainfall

* Corresponding author: wangwanqi@rails.cn

https://doi.org/10.1051/matecconf/202135301007

and construction for comprehensive analysis, dynamically adjusted the construction organization plan, and took effective measures in time, The construction organization plan was dynamically adjusted and effective measures were taken in time to ensure the safety of the project.

The above study discussed the influence of anti-slip piles on slope stability and the influence of anti-slip pile parameter settings on slope stability, but there is no analysis method for the integration of numerical simulation technology and information monitoring technology items.

2 Structural slice model

2.1 Slope slicing model

The slope model can be simulated based on the topographic map of the slope at the site, import the digital map, and then slice the location of the less stable areas for slope stability analysis, so as to determine the location information where anti-slip piles or other measures for reinforcement are needed.

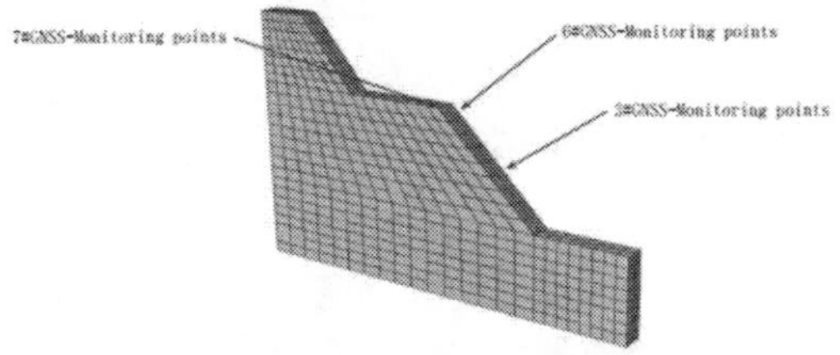

Figure 1 Slope finite element model and monitoring point arrangement

2.2 Slice model under anti-slip pile reinforcement

According to the previously selected reinforcement location, numerical modeling with different reinforcement depth, reinforcement spacing and other parameter variations is carried out, and the specific reinforcement location is determined after considering the economic analysis. After that, slice analysis is performed according to the reinforcement location of the pile to effectively grasp the timely working condition of the anti-slip pile and the reinforcement effect of the soil around the pile.

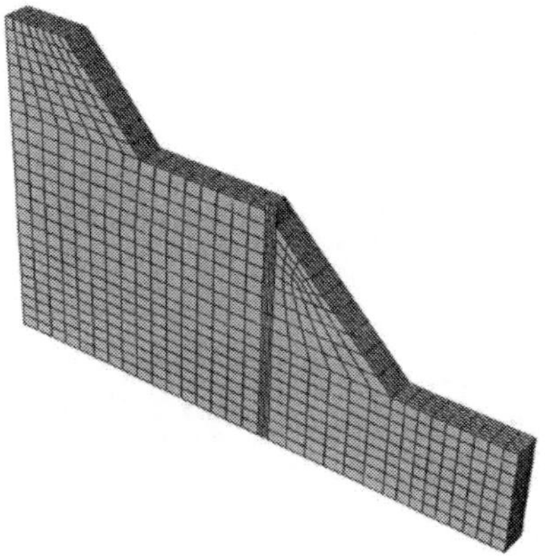

(a) Upper

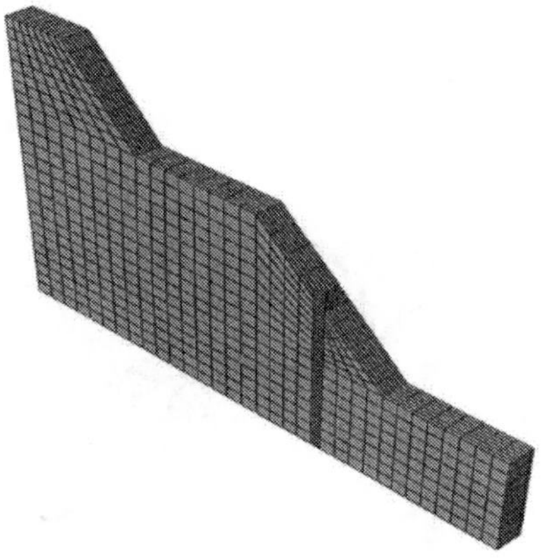

(b) Middle

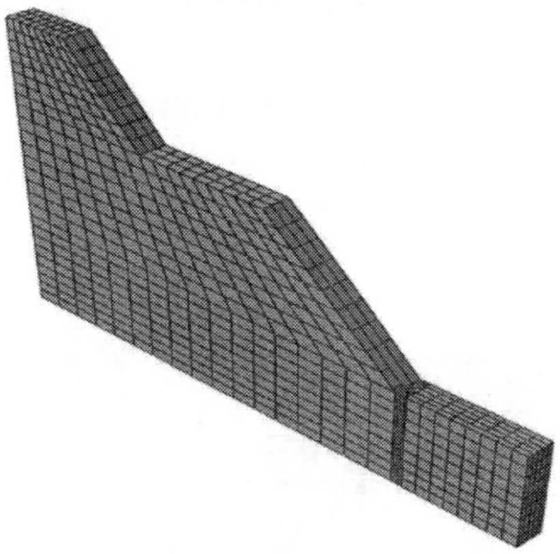

(c) Bottom

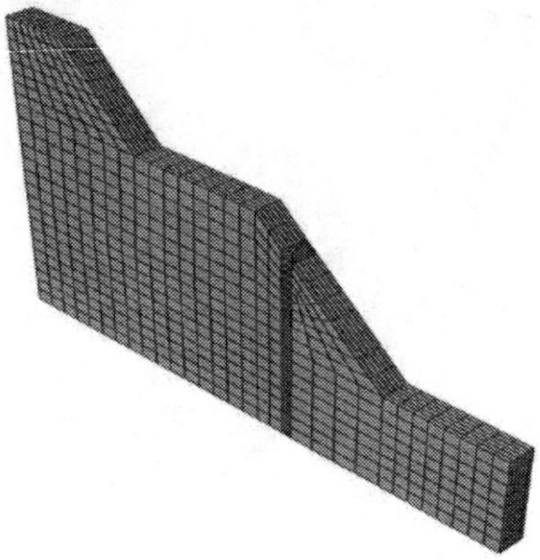

(d) Upper middle

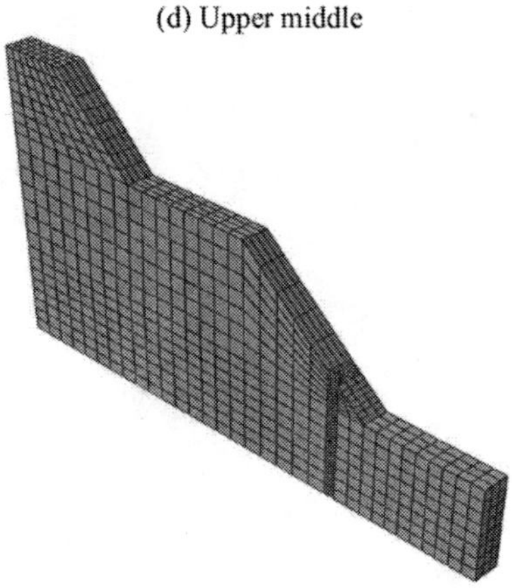

(e) Lower middle

Figure 2 Schematic diagram of anti-slip pile laying

2.3 Numerical simulation results

The displacement clouds of the slope and anti-slip piles when the anti-slip piles are set at the same location as the actual setting will be shown here as a basis for corresponding to the actual monitoring data in the following.

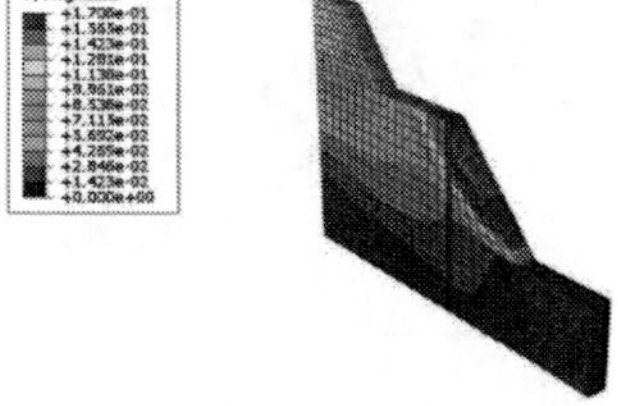

Figure 3 Numerical simulation results (Displacement cloud map)

3 Informative Monitoring

3.1 Information monitoring of the working condition of anti-slip piles

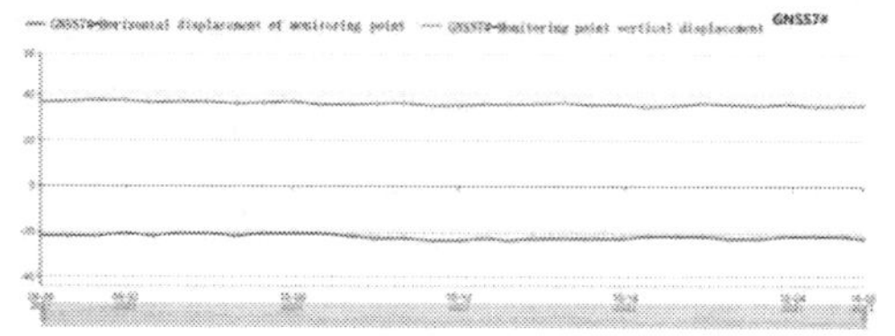

Figure 4 Monitoring point displacement data(GNSS7#)

Point 7 is located on the outside of the anti-slip pile and the upper part of the landslide body, which is used to monitor the displacement of the lower part of the landslide body and the outside of the anti-slip pile.

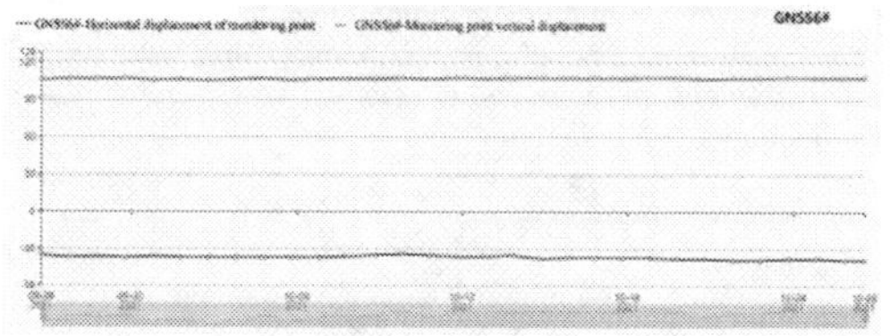

Figure 5 Monitoring point displacement data(GNSS6#)

Point 6 is placed inside the anti-slip pile and the upper part of the landslide body, which is used to monitor the displacement of the upper part of the landslide body and the inner part of the anti-slip pile.
From the monitoring data of GNSS7# and GNSS6#, we can see that the horizontal displacement of 7# is around 37~40 and the vertical displacement is around -18~-23; the horizontal displacement of 6# is around 100~110 and the vertical displacement is around -31~-40. Both monitoring points are in stable condition, and the monitoring data can correspond to the numerical simulation when the anti-slip pile is set in the upper part, as shown in Figure 3.

3.2 Information monitoring of key slope locations

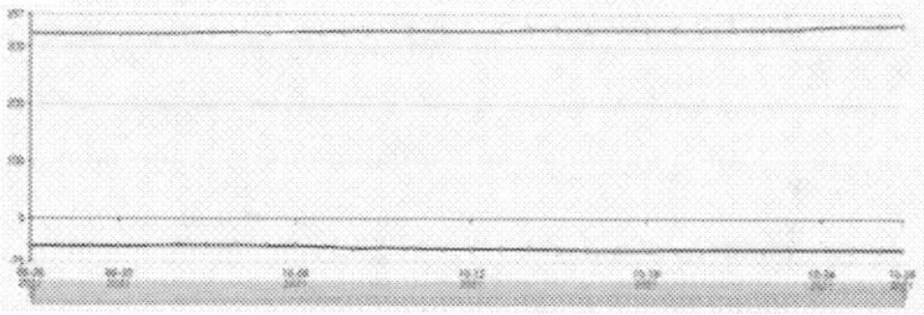

Figure 6 Monitoring point displacement data(GNSS3#)

Point 3 GNSS is placed at the foot of the landslide body to monitor the lower displacement of the landslide body.

4 Conclusions

The piles are monitored by information technology in time, and the reinforcement range is modeled, and an analysis and evaluation method is proposed for the numerical calculation of the slice model and the information technology monitoring data.

References

1. Lei Wenjie, Zheng Yinren, Feng Xiating. Analysis of anti-slip piles piling in landslide management[J]. Geotechnics,2006,27(6):950-954.
2. Zhao Minghua., Liao Binbin., Liu Sisi. Pile spacing calculation of anti-slip piles on slopes based on arch effect[J]. Geotechnics,2010,31(4):1211-1216.
3. Nian Tingkai, Luan Maotian, Zheng Defeng, et al. Limit analysis of the anchorage depth of antislip piles[J]. Journal of Water Resources,2007(6):743-748.
4. Nian Tingkai, Luan Maotian, Zheng Defeng, et al. A simplified analysis method of anti-slip pile considering the effect of pore water pressure in slope[J]. Geotechnics,2008(4):1067-1071.
5. E. Ausilio,E. Conte,G. Dente. Stability analysis of slopes reinforced with piles[J]. Computers and Geotechnics,2001,28(8).
6. HASSIOTIS S,CHAMEAU J L, GUNARATNE M. Design methodfor stabilization of slopes piles[J].Journal of Geotechnical &Geoenvironmental Engineering,1997,125(10):314—323.
7. Jinoh Won,Kwangho You,Sangseom Jeong,Sooil Kim. Coupled effects in stability analysis of pile–slope systems[J]. Computers and Geotechnics,2005,32(4).
8. Dai Zihang. Study on the distribution law of landslide thrust and pile anterior slide resistance of anti-slip piles[J]. Journal of Rock Mechanics and Engineering, 2002, 21(4):517-521.
9. Song Yinjie, Chen Wenqiang, Li Changdong. Progress of slope stability evaluation and pile optimization after anti-slip pile reinforcement[J]. Safety and Environmental Engineering, 2016,23(5):43-49.
10. Liu Bo, Su Peidong, Qiu Peng,et al. Numerical simulation of high-slope anti-slip pile overtopping problem[C].Proceedings of the 2016 National Academic Conference on Engineering Geology.2016.
11. Liu Tao, Zhang Haikuan, Zhang You, et al. Study on the minimum pile spacing of anti-slip piles in three-dimensional composite multilayered landslides[J]. Journal of Rock Mechanics and Engineering, 2018, 37(2):473-484.
12. Zhang Youjun, Wan Champion, Li Yi. Research on the application of multiple monitoring tools in landslide monitoring[J]. Hongshuihe,2010,29(05):26-29.
13. Long Wanxue. Application of monitoring and forecasting technology in landslide information construction [A]. Guizhou Rock Mechanics and Engineering Society, Guiyang Survey and Design Institute of China Hydropower Consultants Group. Proceedings of the 2008 Annual Academic Conference of Guizhou Rock Mechanics and Engineering Society[C]. Guizhou Rock Mechanics and Engineering Society, Guiyang Survey and Design Institute of China Hydropower Consultants Group:Guizhou Science and Technology Association, 2008:4.

Application of New Supercritical Extraction Technology in Extraction and Separation of Effective Components of Traditional Chinese Medicine

Lijuan Cheng

Yichun Vocational Technology College, Yichun 336000, China

Abstract. Supercritical CO2 extraction technology uses supercritical CO2 fluid as extractant to extract and separate effective components from plants from liquid or solid. At present, CO2 is widely used in supercritical fluid, with a critical temperature of 31.265 °C and a critical pressure of 7.18 MPa. The critical conditions are easy to reach. In supercritical state, the density of CO2 fluid is very sensitive to the changes of temperature and pressure, The widely used separation technologies can be divided into two types: membrane separation and traditional separation. The traditional separation methods can be divided into steam distillation, sublimation, cold soaking, precipitation, percolation, decocting method and Soxhlet extraction, etc. These separation methods have certain limitations. In this paper, SFE technology is combined with modern analytical techniques, such as gas chromatography, high performance liquid chromatography, thin layer chromatography, gas chromatography-mass spectrometry, liquid chromatography-mass spectrometry, etc., so that the component analysis can be carried out more efficiently and quickly. Because of the above advantages, SFE is especially beneficial to the extraction of effective components in traditional Chinese medicine and its preparations.

Keywords: New technology of supercritical extraction, Active ingredients of traditional Chinese medicine, Extraction and separation.

1 Introduction

Supercritical CO2 extraction technology is a method that uses CO2 fluid in supercritical state as extractant to extract and separate the effective components in plants from liquid or solid. At present, CO2 is widely used in supercritical fluid, with its critical temperature of 31. 265°C and critical pressure of 7. 18 MPa, and the critical conditions are easy to reach [1]. In supercritical state, the density of CO2 fluid is very sensitive to the change of temperature and pressure, and its dissolving capacity is proportional to pressure within a certain range, so the solubility of the extract can be changed by changing the temperature and pressure of the system, so that it can be separated, thus achieving the purpose of separation and extraction [2]. There are many kinds of natural plants in nature, which contain a variety of effective and complex chemical components. These components can be divided into organic acids, volatile oils, coumarins, steroids, glycosides, alkaloids, sugars, plant pigments, etc. the extraction and separation technology of effective components in plants is based on the existing state, shape Solubility and other physical and chemical properties [3]. At present, the new extraction technology has the characteristics of fast extraction speed, high efficiency, good yield and no pollution, and can retain the original effective components to the maximum extent. Choosing the right extraction method can get twice the result with half the effort. The development of biopharmaceuticals and botanicals has become the trend of innovative drugs in the world. All major pharmaceutical countries in the world are systematically screening effective components from natural plants for developing new drugs. However, China, as the birthplace of botanicals, has hindered the development of traditional Chinese medicines due to insufficient analysis of effective components and backward preparation technology [4].

In addition, SFE technology can also be combined with modern analytical technologies, such as gas chromatography, high performance liquid chromatography, thin layer chromatography, gas chromatography-mass spectrometry, liquid chromatography-mass spectrometry, etc., so that component analysis can be carried out more efficiently and quickly [5]. Therefore, it is widely used in the extraction of effective components of traditional Chinese medicine such as volatile oils, flavonoids, alkaloids, saponins, polysaccharides and quinones [6]. At present, the widely used separation technologies can be divided into two types: membrane separation method and

20175490@qq.com

traditional separation method, and the traditional separation method can be divided into steam distillation method, sublimation method, cold soaking method, precipitation method, percolation method, decocting method method, Soxhlet extraction method, etc. These separation methods have certain limitations [7]. In recent years, the application of new technology in the extraction of traditional Chinese medicine not only conforms to the traditional Chinese medicine theory, but also improves the yield and purity of effective components [8]. This aspect is summarized as follows, in order to provide some reference and guidance for the research and production of modern Chinese medicine pharmacy in China.

2 New technology of extraction, separation and purification

2.1 Supercritical fluid extraction (SFE)

Supercritical fluid extraction technology is a new separation technology rising in the 1960s. The extractant used in this technology is supercritical fluid (SF). There is a critical point for a specific substance. Above the critical point, the substance is in a non gaseous and non liquid state between gas and liquid. The substances within this range are called supercritical fluid (SF). Under the condition of near normal temperature, when the technology is used to extract and separate the effective components in traditional Chinese medicine, it will hardly destroy the effective components in the product; Third, the technology has good selectivity, and the density of supercritical CO2 can be changed by controlling pressure and temperature, so as to change its ability to dissolve substances, and extract some components of Chinese herbal medicine [9]. In the corresponding pressure range, the products obtained by extraction are not single. At this time, the best mixed products can be obtained by controlling the conditions, and the supercritical fluid can be changed into gas by heating or depressurizing, so that all the extracted substances can be precipitated and the separated and purified substances can be obtained. The quality of the extracted products is relatively stable. ③ The extraction ability is strong and the extraction rate is high. ④ The low operating temperature of supercritical fluid carbon dioxide extraction (30~70°C) can better retain the effective components of traditional Chinese medicine. ⑤ No organic solvent is used in the extraction process, so the products produced have no harmful solvent residue. In addition, SFE technology can also be combined with modern analytical technologies, such as thin-layer chromatography, gas chromatography, liquid chromatography, gas chromatography-mass spectrometry, liquid chromatography-mass spectrometry, etc., so it can conduct component analysis more efficiently and quickly. Due to the above advantages, SFE is particularly conducive to the extraction of effective components in traditional Chinese medicine and its preparations [10].

2.2 Ultrasonic extraction technology

Ultrasonic extraction (UE) is a mature and advanced technology applied to the extraction and separation of functional components such as Chinese herbal medicine in recent years. The strong vibration, high acceleration, cavitation effect, thermal effect and stirring produced by ultrasonic can accelerate the entry of plant effective components into the solvent, so as to improve the extraction rate, shorten the time, save the solvent, and avoid the damage of high temperature to the extracted components. The main driving force of ultrasonic extraction is cavitation effect. There are some vacuum or small bubbles containing a small amount of gas in the liquid, When a large number of ultrasonic waves with a certain frequency act on the liquid, small bubbles with appropriate size can produce resonance phenomenon, and they swell rapidly in the sparse stage of acoustic waves. Ultrasonic-assisted extraction mainly improves the leaching speed of effective components of traditional Chinese medicine by means of ultrasonic cavitation. It can improve the extraction efficiency and shorten the extraction time. The mechanical action of ultrasonic can effectively break the cell wall of plants, so that the effective components can be quickly dissolved in the solvent. Ultrasonic extraction of medicinal materials is not limited by the polarity and molecular weight of components, and is suitable for the extraction of most kinds of Chinese medicinal materials and various components. At present, ultrasonic extraction is widely used in small-scale extraction and analysis of traditional Chinese medicine because of its advantages such as economy, convenience and rapidity. Ultrasonic extraction has high extraction efficiency for alkaloids, glycosides, flavonoids and other compounds in traditional Chinese medicine. Ultrasonic extraction technology is a new method for the extraction of plant functional components, which completely changes the traditional extraction process. It is especially suitable for the extraction of effective components of Chinese herbal medicine. Compared with the conventional decoction method, steam distillation method and solvent extraction method, this method has the following characteristics: low extraction temperature, avoiding the adverse effect of long-time heating on effective components, and high biological activity of the product.

3 Application of SFE -CO2 technology in extraction of effective components of traditional Chinese medicine

3.1 Extraction of effective components of volatile oils from traditional Chinese Medicine

Volatile oil is the secondary metabolite in plants. The extraction of volatile oil: A thanasios C K, etc. The volatile oil from garlic was extracted by water distillation, microwave-assisted water distillation and ultrasonic method. It was found that the yield and properties of the volatile oil obtained by the three methods were different, but ultrasonic extraction could reduce the damage to heat-sensitive compounds and was easy to operate. The traditional extraction process of volatile oil is steam distillation, which has the disadvantages of long extraction time, high extraction temperature and easy destruction of effective components, resulting in lower extraction yield of volatile oil with lower boiling point. Under the best extraction conditions, the supercritical CO2 fluid extraction technology was compared with the traditional steam distillation technology. The result was that the volatile oil extracted by supercritical CO2 was nearly 4 times higher than that by traditional steam distillation. Moreover, the operation time is greatly shortened and the process is simplified. Clinical and modern pharmacological studies show that volatile oil has strong biological activities in many aspects. The traditional extraction process of volatile oil is steam distillation. Due to the defects of long extraction time, high extraction temperature and easy to destroy the effective components, the extraction yield of volatile oil with low boiling point is low. Because the volatile oil has good solubility in supercritical CO2 fluid and its molecular weight is small, most volatile oils can be directly extracted with supercritical CO2 fluid [14]. The extraction of volatile oil components of traditional Chinese medicine by supercritical CO2 extraction technology is shown in Table 1.

Table 1 Extraction of volatile oil from traditional Chinese medicine by supercritical CO2 extraction

Process condition	CO2 flow/(l h-1)	Extraction pressure /MPa	Extraction temperature/ °C	Extraction time /h	Extraction rate/%
Mint volatile oil	39	14	49	1.4	3.1
Dried ginger volatile oil	31	26	46	4	4.6
Volatile Oil from Cyperus rotundus and Ligusticum chuanxiong Hort	9	36	46	3	7.65
Amomum cardamom volatile oil	21	27	50	2	2.94

Extraction of volatile oil from zedoary turmeric β-Elemene and the volatile oil extracted from Atractylodes macrocephala are extracted by supercritical CO2 fluid extraction technology under the best conditions or processes. The volatile components of the extract are higher than those of the traditional extract in terms of compound composition and content, which also shows that supercritical CO2 fluid extraction has the advantages of less time-consuming, high efficiency, complete extraction No organic solvent residue.

3.2 Extraction of Alkaloids from Traditional Chinese Medicine

Alkaloids (alkaloid) are a larger class of plant secondary metabolites. At present, there are about 5000 ~ 7000 kinds of alkaloids isolated from plants. Alkaloids are easy to form salts with various acids in plants, and most of them exist in plants in the form of salts, so their solubility in supercritical CO2 fluid is low. When extracting with SFE-CO2, raw materials generally need to be pretreated with alkaline reagents (such as ammonia, triethylamine, Ca (OH) 2, Na2CO3 solution, etc.). Extraction of alkaloids: Yuegang Z et al. In recent years, great progress has been made in the separation of alkaloids from traditional Chinese medicine by supercritical fluid technology, and a large number of valuable alkaloids have been developed. For example, in the extraction of total alkaloids from Leonurus japonicus, the alkalized Leonurus japonicus was extracted by supercritical CO2 after adding entrainer, which can greatly improve the extraction rate of alkaloids, which is 10 times higher than that of conventional methods [15]. The research on SFE and content determination of Aconitum yadong total alkaloids [16] also showed that the yield and content of Aconitum yadong total alkaloids were lower without alkalization treatment, but after alkalization, the effect was significantly improved. See Table 2 for supercritical CO2 extraction of alkaloids from traditional Chinese medicine.

Table 2 examples of supercritical CO2 extraction of alkaloids from traditional Chinese Medicine

Chinese herbal medicine	Main target components	SFE-CO2 conditions	Other extraction methods	References
Datura flower	Scopolamine	At 34. 9 MPa,40°C, 0. 1 m L ammonia water and 0. 2m L methanol as entrainer for 5 min, the impurities are less	Alkaline chloroform extraction, solvent residue	[3]
Chi chi	Piperine	38.5 MPa, 70 °C, 5 min, yield 2.92%	Traditional solvent method is time-consuming and complex in post-treatment	[4]
Light mushroom seed	Colchicine	10 MPa, 45 °C, φ=76% ethanol as entrainer, the yield is 1. 25 times that of solvent method	The yield of traditional solvent method is low	[5]
Cork tree	Berberine and palmatine	Methanol is used as entrainer	Traditional solvent method is time consuming	[6]
Nux vomica	Dening	Ammonia as alkaline agent and acetone as entrainer	Chloroform extraction efficiency is low	[7]
Corydalis tuber	Tetrahydropalmatine	Benzene as entrainer, Ca(OH)2 as alkaline agent, 20 m in	Traditional solvent method is time-consuming and complicated	[8]

 https://doi.org/10.1051/matecconf/202135301008

The ultrasonic method was used to extract nicotine from drugs. It was found that the extraction rate of conventional cold immersion method for 24 hours could be obtained in less than 20 minutes, and the amount of solvent was only 1 / 6 of that of conventional method. There are many kinds of alkaloids, generally referring to the basic nitrogen-containing compounds existing in dicotyledons. Most of them have complex nitrogen-containing heterocyclic structure (nitrogen is mostly contained in the ring). Adding suitable entrainers can also improve the solubility of extracted components, and at the same time, improve the extraction selectivity and increase the purity of extracted components. Commonly used entrainers are mostly methanol, ethanol, acetone, chloroform, etc. Alkaloids have significant physiological effects, such as acting on nervous system, cardiovascular system, anti-inflammatory, antibacterial, antiviral, anti-tumor and antimalarial effects, etc.

4 Conclusions

With the acceleration of the modernization of traditional Chinese medicine, supercritical fluid extraction technology has broad prospects in the extraction and separation of traditional Chinese medicine. It has the characteristics of fast, high extraction rate and less environmental pollution. However, with the all-round development of science and technology, the traditional extraction methods and the above eight newer extraction and separation technologies will develop together, and even more convenient and fast separation technologies will be born in the process of complementarity and intersection. The use of various analytical methods and instruments has a certain scope of application and their advantages and disadvantages. In the analysis process of effective components of traditional Chinese medicine, we should comprehensively consider the specific analysis according to the specific situation to determine the best extraction, separation, analysis and determination method. In the practice of scientific research and production, it is necessary to analyze specific problems in order to select the best extraction method, which should not be blindly selected. In the future, the extraction technology of effective components from natural plants will be safer, more efficient and more environmentally friendly, and the effective components from natural plants will play a role in more fields. Strengthen the application of new technology, study the influence of new technology on the extraction and separation of different drugs, seek the best operating conditions and mechanism, and design the production equipment and process in a targeted manner, so as to realize the leap-forward development of Chinese traditional medicine industry and make greater contributions to human health.

References

1. Wang Wuying. Application of new supercritical extraction technology in the extraction and separation of active ingredients of traditional Chinese medicine[J]. Chinese Medical Guide, 2017, 015(035): 190-191.
2. Liu Na. The application of supercritical fluid extraction technology in the extraction of traditional Chinese medicine[J]. Guangzhou Chemical Industry, 2017, 45(024):31-33.
3. Ding Yuanqing, Wu Xiujuan, Hao Wenyan, et al. Research progress on extraction technology and analysis methods of active ingredients in traditional Chinese medicine [J]. Shandong Chemical Industry, 2017, 46(003): 59-60.
4. Deng Qiaoyu, Jiang Shan, Chen Yudan, Chen Xingguang, Liang Xu, Li Yuzhao, Yuan Jingquan. Application progress of supercritical carbon dioxide extraction technology in the field of traditional Chinese medicine[J]. China Pharmaceuticals, 2020, v.29; No.516(17) :7-11.
5. Deng Peng, Wei Feixue, Wen Wen, et al. Comparative study on extraction of volatile oil from Ligusticum chuanxiong medicinal material by supercritical CO_2 and steam distillation[J]. Journal of Pharmaceutical Practice, 2020, v.38(02):60-63.
6. Chen Yongxing, Chen Yongdong. Analysis of the current situation and technological development of Chinese traditional medicine extraction equipment[J]. Journal of Science & Technology Economics, 2016, 000(004): 75-76.
7. Xie Weiyou[1], Yu Haofei[1], Zhang Lanchun[2], et al. Supercritical CO2 extraction process of the effective ingredients of Chinese medicine Cui Ling Dan [J]. Journal of Kunming Medical University, 2017, 038(005): P.5-9.
8. Quan Chunmei. Two methods of extracting peony essential oil components and their antioxidant activity [J]. Journal of Gansu University of Traditional Chinese Medicine, 2018, 035(006): P.32-36.
9. Wang Haidong, Han Rongxin, Zhang Hongyin, et al. Fractional extraction of Schisandra protein by Osborne method and comparison of antioxidant activity[J]. Food Industry Science and Technology, 2021, 42:1-7.
10. Wei Xiaonan, Hao Tiecheng. Research progress on new technology of Chinese medicine extraction[J]. Chinese Wild Plant Resources, 2020, v.39; No.198(09):51-54.

 https://doi.org/10.1051/matecconf/202135301009

Discussion on the Key Links in the Process of Building Material Quality Inspection and Testing

Zhixin Sun

Civil Engineering and Architecture, Nantong Institute of Technology, Nantong 226002, China

Abstract. With the development of the construction industry, the applied building materials are becoming more and more diversified. As we all know, the performance and quality of the building materials themselves will have a direct impact on the quality of the construction project, which is related to the usability and safety of the construction project. If the quality of building materials cannot be inspected effectively, the probability of safety accidents in construction projects will be greatly increased, which will seriously threaten people's lives and property. This article analyzes the development process of building material testing and testing, and on the basis of summarizing the importance of building material testing and testing, it analyzes several key aspects of building material testing and testing.

Keywords: Building materials;Testing; Inspection.

1 Overview of the development of building materials testing and testing

Architecture is the basic construction of our human society and the most basic needs of people's lives[1]. It is related to people's safe accommodation, production, work, entertainment and rest. With the rapid economic development and the general improvement of living standards, people's requirements for building quality are becoming more and more stringent. The quality of construction involves many factors, among which the most important and direct factor is construction materials. Therefore, inspection and testing of construction materials has become a key link in construction engineering. Compared with the United States, Japan and other developed countries, my country's construction engineering quality inspection started late[2]. Although my country's investment in construction engineering has increased year by year since the reform and opening up, there are still single inspection items, unreasonable inspection sampling, and automated equipment. Problems such as low level and insufficient allocation of professionals are as follows:

The current testing and testing of building materials mainly involves the development of different testing and testing methods based on the physical and chemical properties of the materials. In the specific testing process, in order to ensure the effectiveness of the standardized test results of the quality testing process, it is necessary to conduct a comprehensive test on the function and type of the material[3]. Cement is a common building material and has a wide range of uses in construction. The testing and testing of cement usually include specific gravity and bulk density, fineness and strength, setting time and stability. Different types of cement have different quality standards. If the test items are too single, accurate quality reports cannot be generated. The quality of buildings built on this basis is even more unsafe.

The testing of building materials includes the steps of sampling, sending samples, and witnessing. If the quality of the materials on site is unqualified after inspection, it should be banned immediately upon discovery[4]. Among them, sampling is the process of extracting experimental samples from the inspection objects in accordance with the relevant technical standards and specifications. If the sampling of on-site materials is unreasonable, it is likely to cause missed inspections or re-examinations. The report will not faithfully reflect the quality of building materials, and will increase the cost of testing and cause unnecessary waste of resources.

In all kinds of building construction, the low degree of automation of building materials detection is a common problem[5]. At present, the quality evaluation system of building materials in my country's construction industry is not perfect. In the testing of building materials, many testing items must be completed manually. Compared with the large-scale use of machines for detection, manual detection has problems such as low efficiency and high error rate, which causes deviations in detection results.

The testing technology of building materials involves building materials production, engineering construction, material research and development, quality control and commercial trade. Building materials testing is essential in building construction. Therefore, there is a high

* Corresponding author: szxmsr@163.com

demand for technical testing personnel with high professional quality. However, the current construction industry has the problem of matching high-level professionals with job requirements[6]. The insufficient allocation of inspectors engaged in relevant inspection institutions, quality management personnel, and researchers from construction scientific research units makes the inspection effect impossible to guarantee.

In addition, the testing and testing of building materials still have problems such as non-standard testing technology and imperfect supervision system, which have brought great safety risks to the quality of construction projects.

2 Solutions to existing problems in building material testing

2.1 Material sampling

The material performance test report is obtained by testing the material samples taken. A representative sampling of test materials is essential. Sampling is generally based on randomly selecting a specified number of samples from different parts of a batch of materials, that is, not only the sampling quantity must be correct[7]. The sampling location and method should also be carried out in accordance with regulations. The number of samples is related to the accuracy of the test results. If sampling is not carried out in accordance with the specified sampling method, sampling location and sampling amount, or the sampling situation is deviated from the requirements, it may lead to deviations in the test data, which will adversely affect the quality inspection of the entire material. When receiving the test samples, the staff must do a good job of the state of the samples, and the record must include whether the sample is abnormal or whether it deviates from the standard state described in the test process. The laboratory must first clarify that the preparations for sample inspection are in place, and carry out the preparatory work for sample inspection with reference to the opinions of the client. The laboratory shall specify and appropriate facilities in the quality documents to prevent the samples used for inspection from being damaged or deteriorated before storage, disposal or inspection. The storage conditions must meet the requirements, and sufficient safety measures must be taken to ensure that the samples are inspected. The condition is intact, thus ensuring the smooth development of inspection work. The flow chart of construction material quality inspection management system is shown in Figure 1.

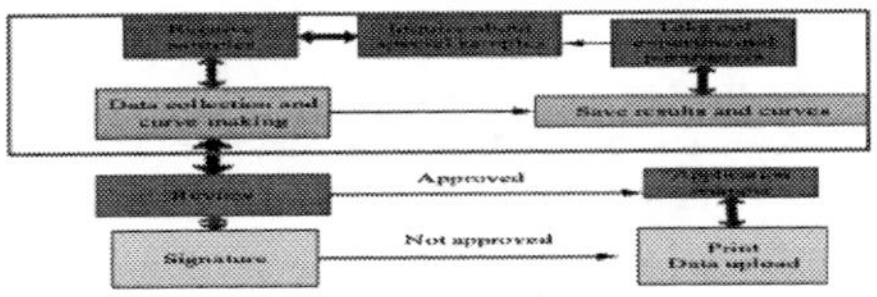

Figure 1. Flow chart of construction material quality inspection management system

What the data acquisition system needs to complete is the "test" step of the mechanical test in the construction material quality inspection management system. Simply put, obtain the detailed content of the commissioned test, perform the test, save the test result in the database, and return to the construction material quality inspection management system.

2.2 Staff quality

Testing agencies must standardize personnel and implement a qualification system. At present, there are a large number of employees in the testing industry, and it is quite difficult to regulate the quality of personnel[8]. It is recommended to further deepen the current training and assessment model of inspectors and implement a practicing qualification system. In addition, sufficient staff should be provided. The laboratory shall clearly stipulate the qualifications of the technical leader and the quality leader, and clearly stipulate the necessary qualifications for the senior and middle management personnel and the qualifications for the technical inspection personnel. It should also clearly specify the qualifications of the supervisory personnel for the verification audit. Qualifications and conditions of the position, and regular training, assessment and recognition.The system function block diagram of data acquisition system is shown in Figure 2.

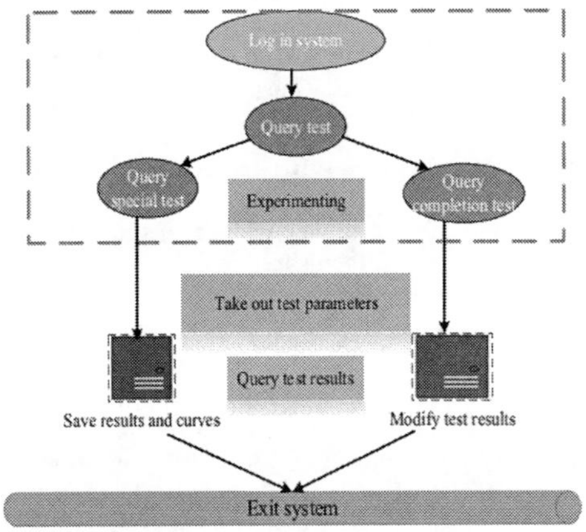

Figure 2. System function block diagram of data acquisition system

2.3 Test environment

Temperature and humidity have a great influence on the performance of some building materials. Therefore, there are clear regulations on material maintenance and environmental conditions during testing in the standard. Only when these regulations are strictly followed can the test results be comparable[9].Under normal temperature test conditions, if the material is applied for mechanical performance test faster, the deformation of the specimen lags behind the load on it, and the measured strength value will be higher than the inherent strength of the material. If the loading speed is faster when measuring the yield point of the steel bar, the yield point value will increase. The bending resistance, compression resistance and loading speed of cement, concrete, brick and other specimens have an impact on the measurement results. Therefore, the loading speed should be operated in strict accordance with the material standards and operating regulations. The loading should be continuous and uniform. When the specimen begins to rapidly deform and approach failure, stop adjusting the throttle of the testing machine until the maximum load value of the specimen is measured.

2.4 Equipment

The laboratory should be properly equipped with all instruments and equipment (including reference materials) for testing. All instruments and equipment are maintained in a timely manner, and maintenance data processing procedures are in place[10]. The testing equipment must be verified and calibrated, and the instruments and equipment should be clearly marked to indicate their status. When the displayed result is suspicious or the verification indicates that there is a defect, it should be stopped immediately and clearly marked. The repaired instrument must be calibrated to prove that it has been restored before it can be used normally to ensure that the data is true, valid and without deviation.In the data acquisition system, after querying the test, double-click a row in the test list to query the test's collection results and modify the data. The modification process will be recorded in the database, and the modification can also be done at the same time See the previous changes.

3 Building material quality inspection

The testing method is the technical instrument for the implementation of the inspection. It is an important resource for the laboratory to carry out inspection services and an indispensable process for the implementation of the inspection work. When the laboratory conducts testing or carries out other related business activities, such as sample preparation, extraction, disposal, storage and transmission; estimation of measurement uncertainty; analysis of inspection data, etc., the correct testing method must be selected according to the relevant testing specifications and workflow. Moreover, international or national general testing specifications and standards must be adopted, and they can be used in work practice only after they have been approved by the laboratory technology supervisor[11]. The standard methods should be traceable. If the testing method or work process adopted is not internationally or domestically regulated, the client's consent must be obtained before testing, and a valid document must be formed to ensure that the report is accepted by the user and the client, and form a valid document confirmed in writing. Update the standards in a timely manner.

3.1 Strengthen management of building materials inspection errors

In the process of testing the quality of building materials, the problem of testing errors is widespread. Therefore, in order to improve the quality of building materials testing, it is very necessary to analyze the causes of testing errors and formulate corresponding testing programs. The problems of material sampling, testing and human operation may cause errors in material testing. In particular, the impact of human operation problems is the most serious. Therefore, the professional capabilities of relevant testing personnel should be continuously strengthened to ensure the standardization of each testing process and avoid human factors affecting the quality of testing. The testing organization itself should Increase investment and regularly update testing equipment to ensure the advanced nature of testing technology. Before testing building materials, the equipment should be debugged according to relevant regulations to ensure the accuracy of testing.

For strength values that are not specifically required, the rounding interval is 1N/mm, when the strength value is 200-1000N/mm, the rounding interval is 5N/mm, and the strength value is >1000N/mm, the rounding interval is 10N/mm[12]. In addition, the rule that no contract cannot be revised during the calculation process should also be followed. For example, the apparent density test of sand "Quality Standards and Inspection Methods of Sand for Ordinary Concrete" stipulates that two tests are required, but the apparent density value calculated each time should not be rounded up. The mantissa of the average value of the result can be rounded to 0, otherwise it will increase the influence of error transmission.

3.2 Establish a dynamic inspection quality evaluation system

Building materials testing agencies should establish a dynamic evaluation system for the quality of testing, and implement it in the testing work, discover the deficiencies in the testing work, so that the staff can effectively adjust the specific process of the testing work, and at the same time The job responsibilities of inspectors are strengthened to enhance the inspection quality of building materials, so as to ensure the quality of construction projects. In addition, it is necessary to solve the problems of the inspection environment, not only to optimize the inspection environment, but also to strengthen the inspection instruments. Maintenance, to ensure the standardization of equipment application and the scientific nature of the calibration process, to ensure the quality of testing.

Due to various reasons, the results of the test data in the same group of specimens are sometimes very discrete. In order to make the test results accurate, the standard stipulates that the test result data of some materials must be selected. For example, in the flexural test of the strength of cement mortar, when one of the three strength values exceeds 10% of the average value, this value needs to be eliminated, and the average value of the other two measured strength values is used as the flexural strength result.

$$E(x) = \sum_{k=1}^{n} x_k p_k \quad (1)$$

$$D(x) = E[(x - E(x)]^2 \quad (2)$$

If two of the measured values exceed ±10% of the average value, the remaining one measured value will be used as the flexural strength result. If all three measured values exceed ±10% of the average value, the test must be repeated. The experimental error value is shown in Figure 3. The calculation of the average strength of the compressive specimens of concrete and mortar has its own selection method. It should be noted that you must not simply add up the data to calculate the problem. The method of rounding off the calculated data is carried out in accordance with GB/T8170, and the mantissa shall be rounded to the nearest single or double method, and the number of digits shall be retained according to the standard. The test result data sometimes appears to be too high or too low than expected, the data in the same set of test pieces are very different, or the performance indicators of the same test piece are contradictory, etc., which requires serious treatment, find out the reason, and retest in time and reinspection.

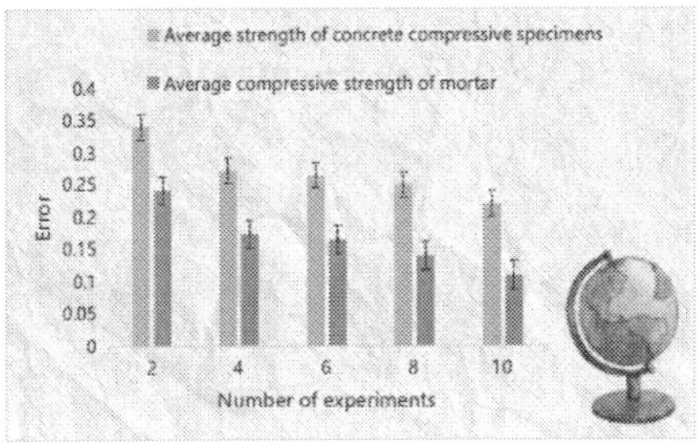

Figure 3. Experimental error value

4 Conclusion

Building material testing is an important link in the quality assurance system of construction projects. Building material testing plays an important role in the development of building materials research and technology, enterprise production and construction engineering construction. It is not only the basis and necessary means for assessing and controlling the quality of building materials It is also an important link for saving raw materials, developing building science and technology, and ensuring project quality. As professional and technical personnel in testing and testing, we must strengthen our own construction, strive to improve work responsibility, summarize experience and lessons in a timely manner, continuously enrich theoretical knowledge, and improve practical operations. We should also strengthen the professional ethics, integrity construction, and discipline of all personnel. A regular assessment system for all types of personnel should be established and improved to ensure the accuracy of test results.

References

1. Nasr M M, Krumme M, Matsuda Y, et al. Regulatory perspectives on continuous pharmaceutical manufacturing: moving from theory to practice: International symposium on the continuous manufacturing of pharmaceuticals[J]. Journal of Pharmaceutical Sciences, 2017, 106(11): 3199-3206.
2. Lian J, Cai O, Dong X, et al. Health monitoring and safety evaluation of the offshore wind turbine structure: a review and discussion of future development[J]. Sustainability, 2019, 11(2): 494-495.
3. Hussain S, Fangwei Z, Siddiqi A F, et al. Structural equation model for evaluating factors affecting quality of social infrastructure projects[J]. Sustainability, 2018, 10(5): 14-15.
4. Cammarota G, Ianiro G, Tilg H, et al. European consensus conference on faecal microbiota transplantation in clinical practice[J]. Gut, 2017, 66(4): 569-580.
5. Attia S, Bilir S, Safy T, et al. Current trends and future challenges in the performance assessment of adaptive façade systems[J]. Energy and Buildings, 2018, 17(9): 165-182.
6. Toledo J B, Arnold M, Kastenmüller G, et al. Metabolic network failures in Alzheimer's disease: A biochemical road map[J]. Alzheimer's & Dementia, 2017, 13(9): 965-984.
7. Pauwels P, de Farias T M, Zhang C, et al. A performance benchmark over semantic rule checking approaches in construction industry[J]. Advanced Engineering Informatics, 2017, 3(3): 68-88.
8. Nicoletto G, Konečná R, et al. Surface roughness and directional fatigue behavior of as-built EBM and DMLS[J]. International Journal of Fatigue, 2018, 11(6): 140-148.
9. Spears T G, Gold S A. In-process sensing in selective laser melting (SLM) additive manufacturing[J]. Integrating Materials and Manufacturing Innovation, 2016, 5(1): 16-40.
10. Rakha T, Gorodetsky A. Review of Unmanned Aerial System (UAS) applications in the built environment: Towards automated building inspection procedures using drones[J]. Automation in Construction, 2018, 9(3): 252-264.
11. Bianchini Ciampoli L, Tosti F, Economou N, et al. Signal processing of GPR data for road surveys[J]. Geosciences, 2019, 9(2): 96-97.
12. Suryanto B H R, Du H L, Wang D, et al. Challenges and prospects in the catalysis of electroreduction of nitrogen to ammonia[J]. Nature Catalysis, 2019, 2(4): 290-296.

Study on the Corrosion Resistance of 39SiCrVTiA High strength and high toughness spring steel

Lun Nie[1,*], Min Zhu [2], Shirun Tu [1], Kefeng Yuan [1], Kexin Lu [1]

[1] Zhejiang Meili High Technology Co., Ltd, Xinchang 312500, China
[2] School of Mechanical Engineering & Automation, Zhejiang Sci-Tech University, Hangzhou 310018, China

Abstract. 39SiCrVTiA spring steel is heat-treated and compared with the existing high-strength spring steels 60Si2CrVA and SAE9254 for electrochemical impedance spectroscopy (EIS), polarization curve and slow strain rate testing (SSRT). The test results of electrochemical impedance spectroscopy (EIS), polarization curve show that the corrosion resistance of 60Si2CrVA was the best, followed by that of SAE9254 and 39SiCrVTIA.However, the test results of the SSRT test show that the three spring steels in 5% NaCl solution possess high SCC susceptibility. The SCC susceptibility of 39SiCrVTiA steel is slightly lower and the stress corrosion ability is better than the other two steels which may be related to its containing Ti, V elements and lower carbon content.

Keywords: 39SiCrVTiA, heat-treated, electrochemical impedance spectroscopy (EIS)

1 Introduction

High strength spring steel is widely used in railway, automobile, engineering machinery and other machinery manufacturing[1].This kind of spring steel, has been implemented in the high-stress design, such as the spiral spring for automobile suspension, but the further improving the strength index of steel has been greatly restricted. With the increase of spring stress, the sensitivity of spring steel to corrosion has increased rapidly, and the problem of corrosion fatigue failure during the process has attracted attention. Therefore, higher strength and better corrosion fatigue resistance will become the main research direction of new spring materials.

According to the current high-performance, high-reliability and lightweight development trends of various types of mechanical devices, this paper develops alloy spring steel with high strength, high toughness and good corrosion resistance. In the research and development of the current high-end spring products such as automotive and rail transit damping spring and stabilizer rod, the latest concept requirements of performance are put forward, enhancing the strength and toughness of the material at the same time pay attention to improve the spring material corrosion resistance to meet the needs of high reliability and lightweight mechanical products.

At present, 55SiCrA and 60Si2CrVAT are used as spring steels. The 55SiCrA steel has unstable strength and poor corrosion resistance, while the 60Si2CrVAT steel has the characteristics of high strength but insufficient plastic toughness, hydrogen embrittlement resistance and corrosion fatigue resistance. Therefore, the research and development of spring steel with high strength, high toughness and corrosion resistance has important practical engineering significance and good academic value. In view of the above situation, in order to improve the low-temperature toughness of the material and ensure the durability of the coil spring in the corrosive environment, the 39SiCrVTiA spring steel with high strength and toughness has been developed.

2 Test Method

This paper mainly uses various testing methods to comprehensively evaluate the various properties of the 39SiCrVTiA spring steel developed, with special attention to corrosion resistance. And compared with the high-strength spring steel 60Si2CrVA and SAE9254 commonly used today, it provides relevant reference basis and data support for the promotion and application of this material. The experimental materials are 39SiCrVTiA, 60Si2CrVA and SAE9254 spring steels with ϕ 15.8 mm. The material is quenched and tempered, and the structure is tempered troostite. The specific chemical components are shown in table 1. 39SiCrVTiA also contains Ni 0.22, Cu 0.23 and Ti 0.049.

Table 1. Chemical compositions of 39SiCrVTiA, 60Si2CrVA and SAE9254 spring steels (mass fraction, %)

Steel	C	Si	Mn	Cr	S	P	V	Fe
39SiCrVTA	0.39	1.75	0.64	0.73	0.003	0.006	0.06	Bal.
60Si2CrVA	0.56	1.56	0.62	1.02	0.009	0.012	0.16	Bal.
SAE9254	0.55	1.50	0.65	0.80	0.009	0.01	--	Bal.

* Corresponding author: lun.nie@china-springs.com

The electrochemical test instrument is a PARSTAT 2273 electrochemical workstation. The sample size is ϕ 15.8 mm × 5 mm. Use a welding gun and a tin wire to solder the back of the test surface of the sample to the copper wire, and seal it with epoxy resin and ethylenediamine. The electrochemical test adopts a three-electrode system, saturated calomel electrode(SCE) as the reference electrode, graphite as the auxiliary electrode, and the sample to be tested is the working electrode. Before the experiment, the working electrode was first polished by SiC water abrasive paper from 150# to 1500#, then cleaned by ethanol and deionized water, and then dried by cold air for later use. The experimental medium is a 5% (mass fraction) NaCl solution at 35 °C. During the experiment, the temperature of the medium is controlled by a constant temperature water bath with digital display. Before the test, the sample is polarized at -1000 mV for 3 minutes. The electrochemical impedance spectroscopy (EIS) test frequency range is 105 ~ 10-2 Hz, and the excitation signal amplitude is 10 mV. The EIS test is performed under open circuit potential. The impedance data is shown by Nyquist and Bode diagram. The circuit component codes were used to describe the data, and the equivalent circuit was fitted and analyzed with ZSLMPWLN software. When measuring the polarization curve, the potential scanning range is: -1000~-450 mV (vs SCE), and the scanning rate is 0.5 mV·s-1. The polarization curve results are analyzed by Tafel extrapolation.

The slow strain rate tensile test (SSRT) is carried out on the WDML-30KN material testing machine. The experiment uses a smooth plate-like tensile test specimen with a sample size prepared in accordance with GB/T 15970. Before the experiment, the surface of the sample is polished to 1500 # by SiC water-resistant sandpaper, then cleaned with acetone, deionized water, and dried for later use. The experimental solution medium is 5%NaCl (mass fraction). A three-electrode system was used, with the tensile sample as the working electrode, Pt sheet as the auxiliary electrode, and saturated calomel electrode (SCE) as the reference electrode. In SSRT experiment, the tensile samples are tested at the constant potential of -1000 mV (vs SCE). The test samples were maintained at this potential until the samples fractured during the whole experiment, and the CS350H electrochemical workstation is used to apply constant potential to the samples. The tensile strain rate is 1× 10-6 s-1, and the test temperature is room temperature (~25 °C). After the SSRT experiment, Quanta250 scanning electron microscope (SEM) is used to observe and analyze the fracture morphology of tensile samples.

After the experiment, the parameters of elongation loss Iδ and section shrinkage loss Iψ are used to evaluate the SCC sensitivity of spring steel, where Iδ and Iψ are calculated according to the following formula.

$$I_\delta = \left(1 - \frac{\delta_E}{\delta_0}\right) \times 100\% \quad (1)$$

$$I_\psi = \left(1 - \frac{\psi_E}{\psi_0}\right) \times 100\% \quad (2)$$

In the formula, δE and ψE are the elongation and sectional shrinkage of spring steel in solution, and δ0 and ψ0 are the elongation and sectional shrinkage of spring steel in air, respectively.

3 Results and Analysis

3.1 Electrochemical test.

Electrochemical impedance spectra test. Figure1 shows the electrochemical impedance spectra of three materials in a 5% NaCl solution. The equivalent circuit Rs (QdlRt) in Figure2 is used to fit the impedance spectrum, where Rs is the solution resistance, Qdl is the electric double layer capacitance at the non-ideal electrode/solution interface, and Rt is the charge transfer resistance. In the actual electrochemical system, the frequency response characteristics of the double-layer capacitor at the electrode/solution interface are different from that of the pure capacitor, so the constant phase angle Q is often used to represent the double-layer capacitor. Figure3 shows the fitted charge transfer resistance Rt. It can be seen from Figure3 that the value of the charge transfer resistance Rt of different materials is different. 39SiCrVTiA has the smallest Rt value, and the other two materials are closer. The smaller Rt is, the smaller the resistance during charge transfer is[2], and the lager the electrochemical reaction rate is, the lager the corrosion rate is[3]. Therefore, the order of the corrosion resistance of different materials is: 60Si2CrVA> SAE9254> 39SiCrVTiA.

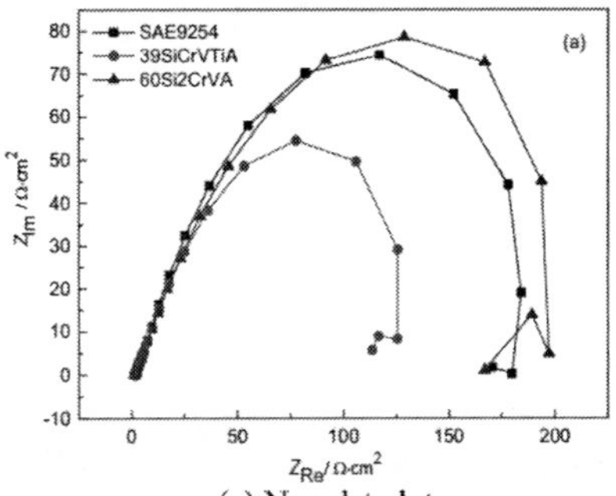

(a) Nyqulst plot

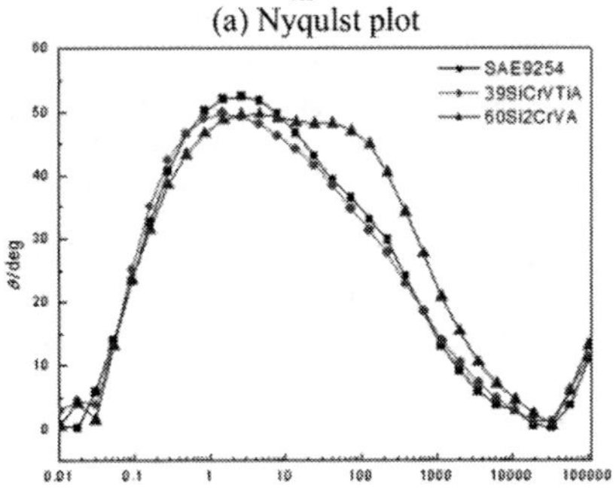

(b) Bode plot of phase angle

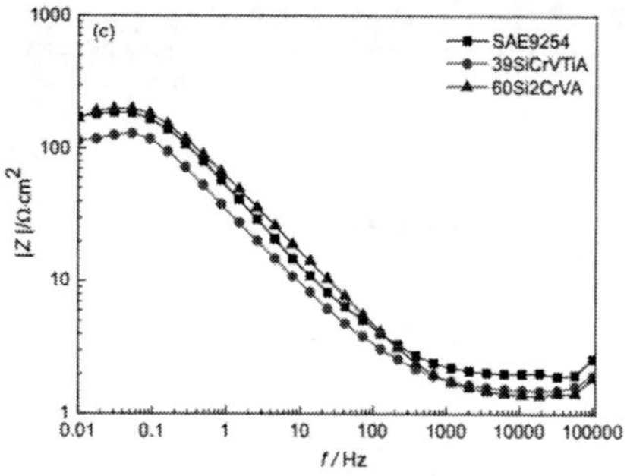

(c) Bode plot of |Z|

Figure 1. Electrochemical impedance spectra of different materials in 5% NaCl solution

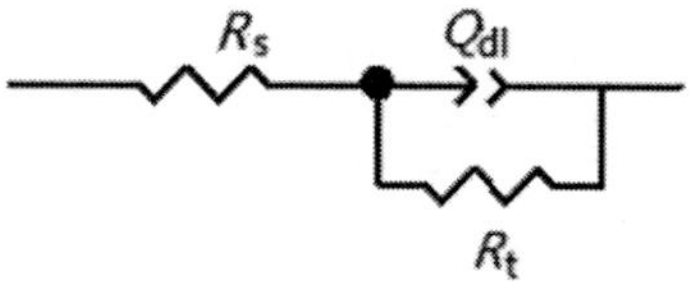

Figure 2. Electrochemical impedance spectroscopy simulation equivalent circuit

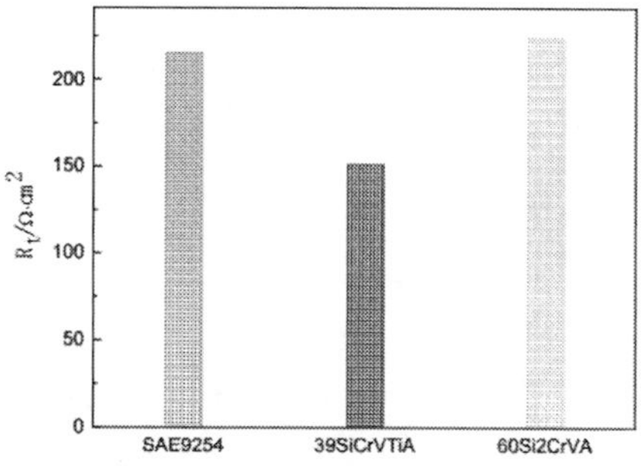

Figure 3. Fitted charge transfer resistance R_t

Polarization curves test. Figure 4 shows the polarization curves of different spring steels in 5% NaCl solution. Table 2 shows the corrosion potential Ecorr and corrosion current density Icorr of different spring steels in 5% NaCl solution fitting from the polarization curves in Figure5. From the table 2, it can be seen that the corrosion potential sequence from positive to negative is 60Si2CrVA, SAE9254 and 39SiCrVTiA, and the corrosion current density sequence from small to large is 60Si2CrVA, SAE9254 and 39SiCrVTiA.The more negative the corrosion potential is, the greater the thermodynamic tendency of corrosion is. The larger the corrosion current density is, the higher the corrosion rate is [4,5]. Therefore, the corrosion resistance order of different materials is 60Si2CrVA > SAE9254 > 39SiCrVTiA.

Based on the above test results, spring steel 60Si2CrVA has the best corrosion resistance, while the steel SAE9254 and 39SiCrVTiA has poor corrosion resistance. According to the chemical composition of the three spring steels, the composition of steel 39SiCrVTiA contains Cu, Ti, Ni and other elements. Relevant studies have shown that the addition of these three elements in material can improve the electrode potential [6-9], refine the grain and improve the compactness of the rust layer thus material corrosion resistance can be improved. Additionally, the reduction of carbon content in steel 39SiCrVTiA is also beneficial to improving its corrosion resistance. However, the actual test results show that the corrosion resistance of steel 39SiCrVTiA is worse than that of steel 60Si2CrVA. This may be related to the content of the other two elements. According to the chemical composition comparison, steel 60Si2CrVA has higher Cr and V content (1.02% and 0.16%), while the Cr and V content of steel 39SiCrVTiA is only 0.73% and 0.06% respectively. As is known to all, element Cr is the key element to improving the corrosion resistance of materials [10,11], and its content greatly affects the corrosion resistance performance. Meanwhile the element V is useful for refining grains and reducing segregation, thus improving corrosion resistance to a certain extent. Therefore, the higher content of element Cr and V may account for the better corrosion resistance of spring steel 60Si2CrVA.

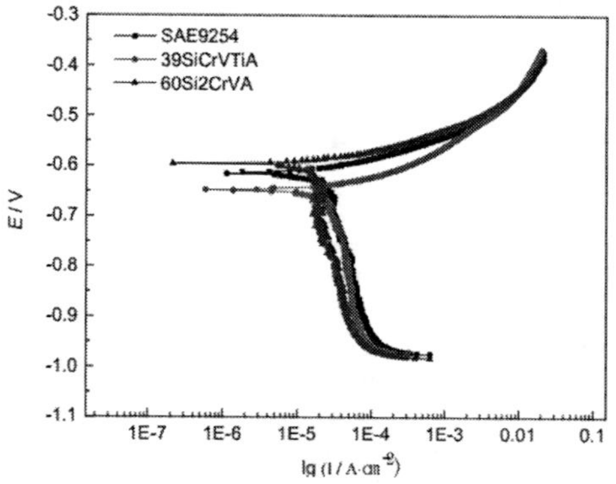

Figure 4. Polarization curves of different spring steels in 5% NaCl solution

Table 2. Corrosion potential Ecorr and corrosion current density Icorr of three steels

Steel	Ecorr /mV(vs SCE)	Icorr /(μA·cm^{-2})
39SiCrVTiA	-651	29.8
60Si2CrVA	-598	17.1
SAE9254	-614	23.1

Stress corrosion test. Figure 5 shows the stress-strain curves of three spring steels in air and 5%NaCl solution. According to the figure, the tensile strength and yield strength 0.2 of spring steel 39SiCrVTiA, 60Si2CrVA and SAE9254 is 1855Mpa and 1767Mpa, 2127Mpa and 1900 Mpa, 2073 Mpa and 1845 Mpa respectively. It can be observed that the stress-strain behavior of samples in air and solution (applying -1v vs.SCE) is significantly different, and the elongation of samples in air is much higher than that in solution. When we carried on the slow strain rate tensile test (SSRT) of the three materials in

5%NaCl solution, the samples fracture during the elastic deformation phase. This indicates that spring steels have a high stress corrosive cracking sensitivity under corrosion test condition.

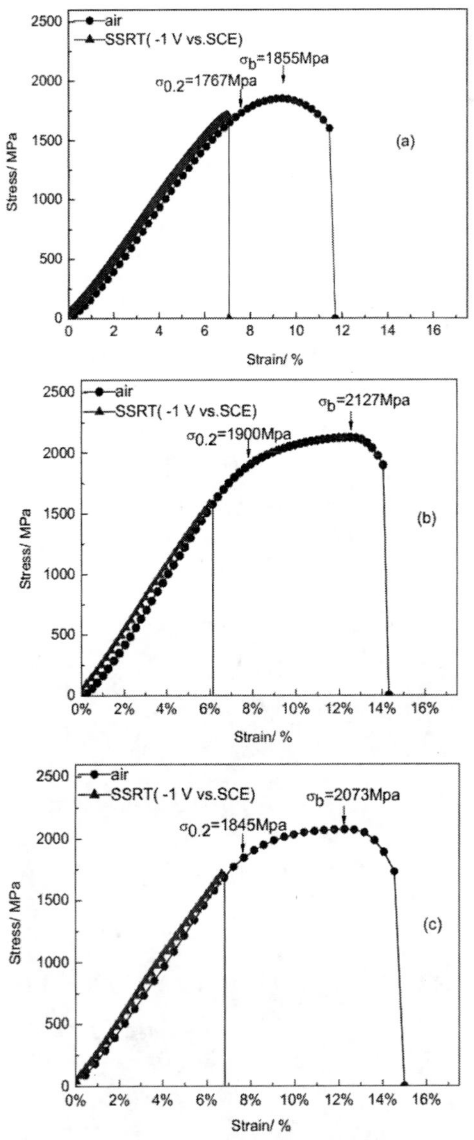

Figure 5. Stress-strain curves of three steels tested in air and 5%NaCl solution (a) 39SiCrVTiA; (b) 60Si2CrVA; (c) SAE9254.

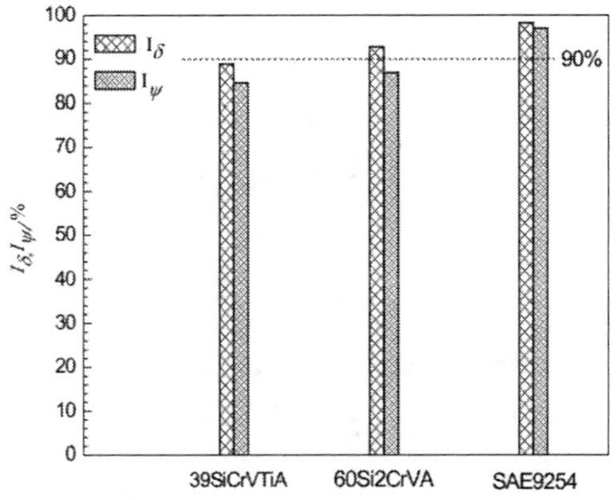

Figure 6. Elongation-loss rate Iδ and area reduction loss rate Iψ of the three spring steels in 5% NaCl solution

Figure 6 shows the elongation loss and section shrinkage loss of three spring steels in 5% NaCl solution. From the figure, the spring steels 39SiCrVTiA and 60Si2CrVA in solution both have the very high elongation loss Iδ and area reduction loss Iψ, while the steel SAE9254 has higher Iδand Iψ that larger than 90%, and SCC parameter values are the lowest. Combining with the Figure8, we can know that three spring steels all have higher SCC sensitivity under corrosion test condition, and steel 39SiCrVTiA has a slightly lower SCC sensitivity than the other spring steels. The high brittleness coefficient indicates that three spring steels brittle fracture.

Figure7 shows the fracture morphologies of spring steel 39SiCrVTiA in air and 5% NaCl solution. It can be seen from Figure7(a)that the tensile fracture of steel 39SiCrVTiA in air has obvious necking and plastic deformation phenomenon. meanwhile. Figure7(b) shows a large number of dimples and micropores of different sizes which shows evident ductile fracture characteristics. Figure7(c) and (d) show that the fracture morphology of spring steel in 5% NaCl solution is different from that in air. Compared with the fracture morphology in air, the fracture is more flat and there is no apparent necking phenomenon when it fractured in 5%NaCl solution. Figure7(d) presents the typical intergranular crack morphology which looks like rock candy. The cracks extand along the grain boundary, and the grain separation phenomenon also can be observed, which are characteristics of Intergranular Stress Corrosion Crack(IGSCC) [12,13]. After observing of Figure7(e) in detail, it can be seen that the cracks in the local region are wide and deep, and the short cracks are connected with each other to grow into long cracks, and the extension path is similar to the linear form, which are the obvious hydrogen embrittlement cracking characteristics (HE)[14]. Therefore, the SCC process presents the characteristics of intergranular brittle cracking and hydrogen embrittlement cracking at the same time, and its SCC sensitivity is relatively high.

 https://doi.org/10.1051/matecconf/202135301010

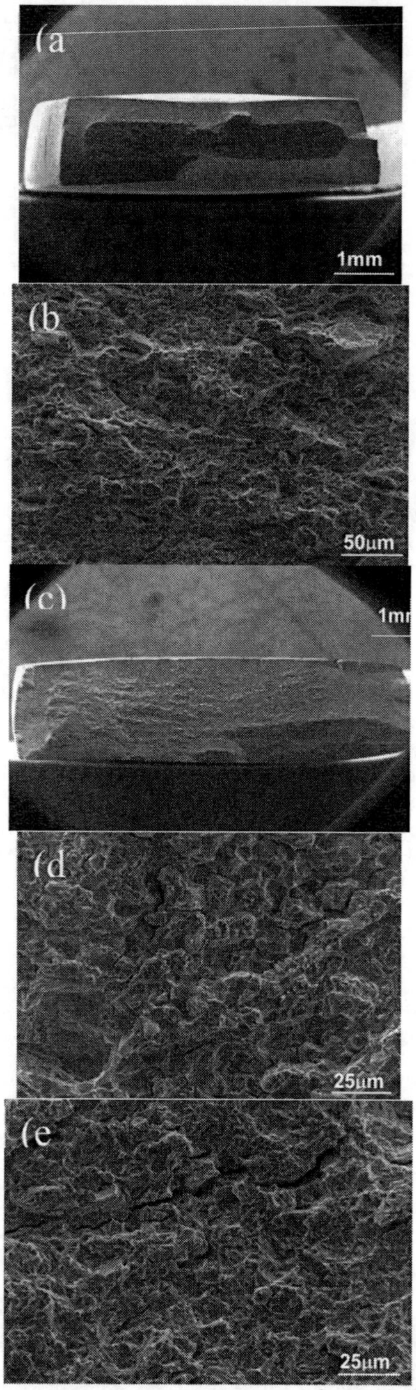

Figure 7. Fracture morphologies of steel 39SiCrVTiA tested in air (a,b)and 5% NaCl solution(c-f)

Figure 8 shows the fracture morphology of spring steel 60Si2CrVA in air and 5% NaCl solution. It can be seen from Figure 8(a) and(b) that the tensile fracture of steel 60Si2CrVA in air also has obvious necking and plastic deformation phenomenon, and large number of dimples and micropores of different sizes and depth which shows ductile fracture characteristics. Figure8(c) and (d) show that the fracture morphology of steel 60Si2CrVA in 5%NaCl solution is significantly different from that in air. The fracture in air shows obvious ductile fracture characteristics like necking phenomenon; however, in 5%NaCl solution, the fracture shows no necking phenomenon. Figure8(d) shows the typical ice-sugar-like intergranular cracking morphology, with many cracks extending along grain boundaries and obvious grain separation, which are obvious characteristics of Intergranular Stress Corrosion Crack (IGSCC). After observing the Figure8(e) and 8(f) in detail, it also can be seen that apart from obvious characteristics of IGSCC , there are also existing some characteristics of Transgranular Stress Corrosion Cracking (TGSCC)[15,16,17] in the local region. Transgranular cracks are long, deep and wide when they almost extend in a continuous linear form, which show the obvious hydrogen embrittlement cracking characteristics (HE).Then, if combined with Figure7 and 8, it can be seen that when -1v (vs.SCE) is applied in SSRT process, the spring steel 60Si2CrVA has a high SCC sensitivity in 5%NaCl solution, and presents obvious characteristics of intergranular brittle cracking and hydrogen embrittlement cracking.

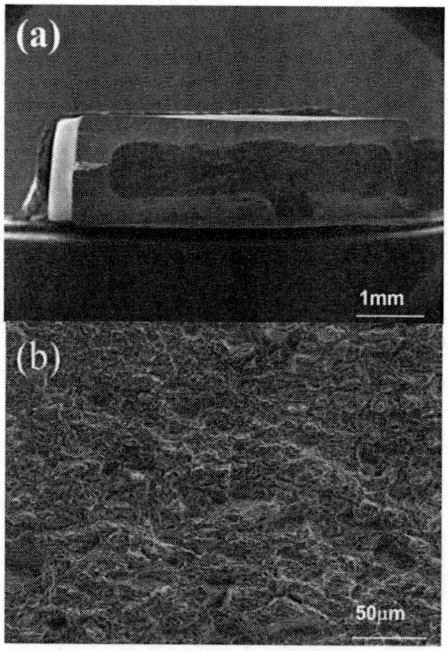

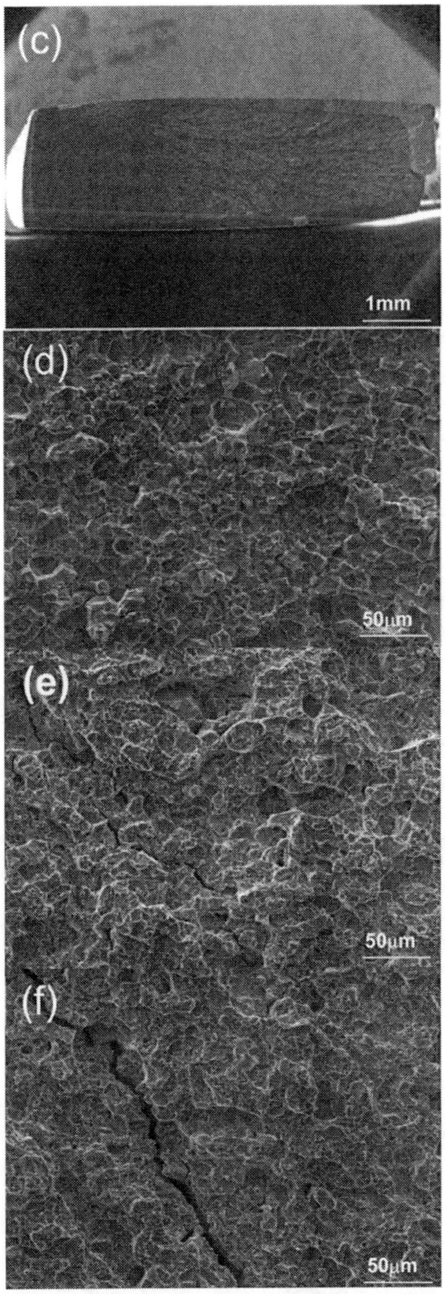

Figure 8. Fracture morphologies of steel 60Si2CrVA tested in air (a,b) and 5% NaCl solution(c-f)

Figure 9 shows the fracture morphology of spring steel SAE9254 in air and 5%NaCl solution. It can be seen from Figure9 (a) and (b) that it is similar to the steel 60Si2CrVA , the steel SAE9254 fracture in the air also presents necking and plastic deformation phenomenon, as well as a large number of dimples and micropores, which on behalf of the ductile fracture. Figure9(c) and (d) show that the fracture of SAE9254 steel in 5% NaCl solution is flat without necking phenomenon , and there is obvious grain separation, which are typical characteristics of hydrogen embrittlement. In addition, Figure9 (e) also shows that long, deep and wide transgranular cracks extending in a straight line. This indicates that the fracture characteristics of steel SAE9254 both have the characteristics of intergranular brittle cracking and transgranular cracking, in the form of mixture of shallow intergranular crack extention and deep transgranular crack extention. It is shown that steel SAE9254 has higher Iδ and Iψ value, and presents the obvious characteristics of hydrogen embrittlement cracking combined with Figure8. So when we apply 1V (vs. SCE) in the process of the SSRT, the steel SAE9254 shows more SCC sensitivity in 5% NaCl solution.

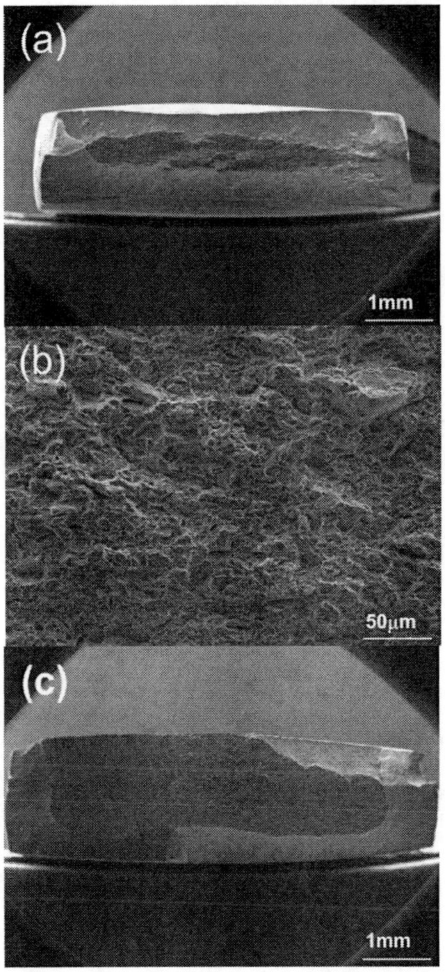

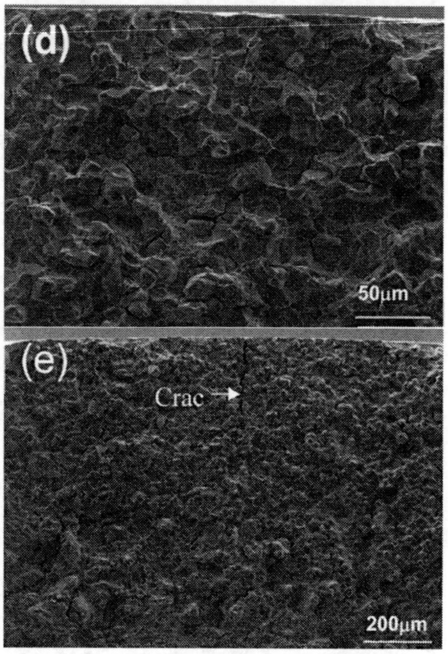

Figure 9. Fracture morphologies of steel SAE9254 tested in air (a,b) and 5% NaCl solution (c-e)

3.2 Discuss

According to the polarization curve (Figure 4), when the potential reaches -1V (vs, the steels will react as the cathode as shown below.

$$2H_2O+2e^- \rightarrow 2H+2OH^- \quad (3)$$

$$H+H \rightarrow H2 \quad (4)$$

The partly H generated in the process of cathode hydrogen evolution will be adsorbed, then it penetrates into the steel and spreads along the grain boundary. In the SSRT process, on the influence of stress, the diffusion of H atom is intensified, and it tends to gather at the grain defect (grain boundary), which will lead to stress concentration and generate a large number of tiny grain boundary cracks in the steel (Figure7(d), 8(d) and 9(d)).In addition, due to the stress-induced diffusion, H atom is more likely to diffuse and accumulate at the crack tip, and it will promote the rapid propagation of the crack tip, then the tendency of hydrogen embrittlement increases and it will lead to the occurrence of transgranular brittle fracture (Figure7 (e), 8(e), 8(f) and 9(e)). If we combine the brittle fracture characteristics in Figure7, 8 and 9 with the relatively large values of section shrinkage loss and elongation loss in Figure6, it can be inferred that when spring steel is applied to -1V (vs.SCE) in the 5% NaCl solution, the SCC mechanism is mainly reflected as hydrogen embrittlement mechanism (HE).

Additionally, it can be seen from table 1 that the main difference in chemical composition of spring steel 60Si2CrVA and SAE9254 is that steel 60Si2CrVA contains V element, while steel SAE9254 does not. In addition to refining grains, reducing segregation and improving strength and toughness, the element V is also the element forming the strong carbide, stable tiny precipitates which avail to the formation of hydrogen capture traps and then affect the diffusion and distribution of hydrogen in steel leading to the effectively reducing the sensitivity of delayed fracture of steel[18,19]. This may be the reason for the higher SCC sensitivity of steel SAE9254 under the experimental conditions in this paper. While spring steel 39SiCrVTiA contains element Ti, which has a similar function to V. The addition of Ti can increase the number of hydrogen traps and improve toughness, thus improving its resistance to hydrogen brittleness. Additionally, the steel 39SiCrVTiA has the lowest carbon content, which is also conducive to improving the hydrogen brittleness resistance. Therefore, the SCC sensitivity of 39SiCrVTiA steel is lower than that of the other two materials.

5 Conclusion

The results of the long period immersion test show that the corrosion rates of the three spring steels are similar, and the steel SAE9254 has the highest corrosion rate, followed by 39SiCrVTiA and 60Si2CrVA steels. Electrochemical ac impedance spectrum and polarization curve tests all show that the corrosion resistance sequence of spring steel materials from strong to weak was 60Si2CrVA > SAE9254 > 39SiCrVTiA.

When -1V (vs.SCE) was applied in SSRT process, all three spring steels had high SCC sensitivity in 5%NaCl solution. The SCC sensitivity of steel 39SiCrVTiA was slightly lower, while the SCC sensitivity of steel SAE9254 was largest. The SCC mechanism is mainly hydrogen embrittlement (HE). The lower SCC sensitivity of 39SiCrVTiA steel may be related to its elements Ti, V and low carbon content.

To sum up, 39SiCrVTiA spring steel is a kind of high strength and high toughness spring steel, which has excellent plastic toughness especially under low temperature condition and it also has the better hydrogen embrittlement resistance. However, its corrosion resistance remains to be improved.

References

1. Ouyang Zhengrong, Zhai Zhenglong, Zhu Jingjun. A brief analysis of the development status and technical points of spring steel[J]. Laigang Science & Technology, 2013,(6):7-10.
2. Ouyang W Z. Equivalent circuit for EIS study on rusted marine cast iron artifacts [J]. Advance Materials Research, 2015, 1094: 316-319.
3. West W C, Hood Z D, Adhikari S P, et al. Reduction of charge-transfer resistance at the solid electrolyte - electrode interface by pulsed laser deposition of films from a crystalline Li2PO2N source [J]. Journal of Power Sources, 2016, 312: 116-122.
4. Liu Shuan,Zhao Xia,Chen Changwei. Corrosion Behavior of Pipeline Steel X65 in Oilfield[J]. Journal of Chinese Society for Corrosion and Protection, 2015,35(5):393-399.

5. Yin Z F, Zhao W Z, Lai W Y, et al. Electrochemical behaviour of Ni-base alloys exposed under oil/gas field environments [J]. Corrosion Science, 2009, 51(8): 1702-1706.
6. Xu Dexian,Yin Zhongda. High Strengthening Spring Steel and Effect of Alloying Elements [J]. Heat Treatment of Meatails, 2003,28 (12): 30-36.
7. Yu Fuzhou. Corrosion resistance of Metal Materials [M]. Bei Jing:Science Press,1982.
8. Xing Xianqiang. Present situation of low-alloy spring steel wire rod for domestic automobile [J]. Metal Products, 2010,(6):34-37.
9. Luo Xiong,Lin zhenyuan. The development of high strength corrosion resistant alloy material in the lightweight design of car suspension spring[C]. Wu Xi: Proceedings of the 23rd annual meeting of national metal products information network, 2013.
10. Huang Jianzhong, Zuo Yu. Corrosion resistance and corrosion data of materials [M]. Bei Jing: Chemical industry press, 2003, 81-84.
11. Hu Lihua, Lu MinXu, Chang Wei. Effect of Cr Content and Microstructures on the CO2 Corrosion Resistance of the Weld Joint of 1%Cr Containing Pipeline Steel [J].Journal of Materials Engineering, 2010, (7): 82-86.
12. Wang J Q, Atrens A. SCC initiation for X65 pipeline steel in the "high" pH carbonate/ bicarbonate solution [J]. Corrosion Science, 2003, 45(10): 2199-2217.
13. Li J, Elboujdaini M, Fang B. Microscopy study intergranular stress corrosion cracking of X-52 line pipe steel [J]. Corrosion, 2006, 62(4): 316-322.
14. Chen W, King F, Vokes E. Characteristics of near-neutral-pH stress corrosion cracks in an X-65 pipeline [J]. Corrosion, 2002, 58(3): 267-275.
15. Guo Hao, Li Guangfu, Cai Xun. Stress Corrosion Cracking Behavior of X70 Pipeline Steel in Near-Neutral Ph Solutions at Different Temperatures [J]. ACTA METALLURGICA Sinice, 2004, 40(9): 967-971.
16. Dmytrakh I M. Corrosion fracture of structural metallic materials: effect of electrochemical conditions in crack [J]. Strain, 2011, 47(s2): 427-435
17. Tang X, Cheng Y F.Quantitative characterization by micro-electrochemical measurements of the synergism of hydrogen,stress and dissolution on near-neutral pH stress corrosion cracking of pipelines [J]. Corrosion Science, 2011, 53(9): 2927-2933.
18. Hui W J, Dong H, Weng Y Q, et al. Delayed fracture behavior of CrMo type high strength steel containing titanium [J]. Journal of Iron and Steel Research Inter, 2005, 12(1):43-49.
19. Xian Aiping, Li Peiji, Li Xun. Trapping of Hydrogen by Precipitate TiC in Low Carbon Steel[J].ACTA METALLURGICA Sinice, 1986, 22(3): A181-187.

Lewis Acid Combined with Microwave Converts Hemicellulose and Cellulose in Cane Pith into 5-Hydroxymethyl furfural and Glucose

Kun du *, Yufeng Zeng and Ronghuan Qin

College of Chemistry and Food Science, Yulin Normal University, Yulin, Guangxi 537000, China

Abstract: The paper using sugarcane pith as a raw material, the feasibility of aluminum hydroxide Lewis acid combined with microwave conditions to be used as a resource was investigated.Under the conditions of the combination of stellate acid and microwave, the 5-hydroxymethyl furfural (5-HMF) in 0.9% of the liquid phase product can be detected, and its low yield of 0.9% is worthy of further improvement. Economic analysis shows that the added value of 5-HMF produced by Lewis acid combined with microwave is as high as about 30,000 yuan per ton, and the prospect of resource utilization is huge. Use oxalic acid to hydrolyze the remaining cellulose and hemicellulose after 5-HMF is prepared from sugarcane pith to glucose, which is used in food and related products. It has very important application value in the industry. In this paper, the hemicellulose in cane pith is converted to 5-HMF under microwave reaction conditions, and then the remaining residue of the reaction is hydrolyzed with oxalic acid to produce glucose, which provides theoretical basis and technical support for further research on the high-value utilization of sugar cane pith in the future.

Key words: Aluminum hydroxide Lewis acid; 5-hydroxymethyl furfural; Oxalic acid; Glucose.

1 Introduction

Bagasse is a by-product of the sugarcane sugar industry, a rich and cheap renewable resource. Bagasse has now become one of the main raw materials for papermaking, especially in recent years Guangxi has vigorously developed bagasse papermaking, and the output of bagasse pulp has been increasing year by year. Major sugar mills have established their own bagasse pulp and paper mills to implement the strategy of combining sugar and paper to improve the economic benefits of enterprises. Research and analysis have shown that bagasse contains 32 ~ 48% cellulose, 19 ~ 24% hemicellulose, and 23 ~ 32%. Lignin, and about 4% ash.

When using bagasse to make paper, it is necessary to separate the sugarcane pith that affects the quality of the paper. Guangxi alone produces 2.5 to 3 million tons of sugarcane pith a year. The composition of sugarcane pith is similar to that of bagasse. The main components are cellulose 32%-40%, hemicellulose 24%-29%, and lignin 27%-32%. It is one of the few plant fibers with extremely high hemicellulose content. The main component of hemicellulose is polypentose, which is easily hydrolyzed into monosaccharides such as xylose, L-arabinose and glucose by dilute acid. At present, the processing of sugarcane pith in Guangxi Sugar Factory is basically burned as fuel or discarded directly. It not only causes environmental pollution but also makes the added value of sugarcane pith very low.and the added value is very low. With the rapid growth of fossil energy consumption and the depletion of its reserves, the development and application of renewable clean energy has attracted much attention. In the report published in 2004, 12 platform molecules of biomass sugars were put forward for the first time, and the "concept of biomass refining" was emphasized. Among them, the awakening of the sleeping giant 5-hydroxymethylfurfural (5-HMF) brings furan compounds and polymers back to the forefront of basic research and industry. After Leshko et al. [1] published an article on the efficient production of 5-hydroxymethylfurfural in a two-phase organic reaction system using fructose as a substrate in Science, 5-HMF formally entered people's attention. Using 5-HMF as a raw material can produce a series of downstream products. 5-HMF is prepared by removing three molecules of water from monosaccharides through catalytic reactions [2]. Disaccharides and polysaccharides such as sucrose, cellobiose, cellulose, etc. can also be used as raw materials to prepare 5-HMF, but firstly, it must undergo a hydrolysis process to depolymerize it into monosaccharides. Although the method seems simple, 5-HMF is easy to rehydrate levulinic acid and formic acid during the reaction, and cross-polymerize into soluble polymers and insoluble humic substances, which

* Corresponding author: huatangchun769@126.com

 https://doi.org/10.1051/matecconf/202135301011

complicates the synthesis process of 5-HMF [3]. At present, researchers have proposed a variety of reaction mechanisms for the dehydration of hexose to produce 5-HMF based on extensive research, which can be roughly divided into pathway 1 and pathway 2, as shown in Figure 1.

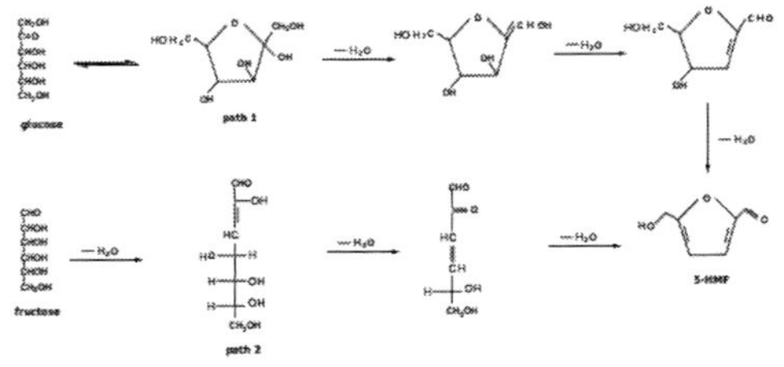

Figure 1. The mechanism of dehydration of glucose and fructose to produce 5-HMF.

It can be seen from Figure 1 that the biggest difference between the two routes lies in whether or not intermediates of cyclic compounds are generated during the conversion of monosaccharides to 5-HMF. In Path 1, glucose is first converted to fructose through isomerization, and then fructose can generate specific intermediates under certain conditions, and further dehydration to generate 5-HMF, while path B directly obtains 5-HMF by removing three molecules of water. At present, more research results indicate that the main mechanism of hexose dehydration to produce 5-HMF is path 1. Amarasekara et al. [4] used ^{1}H-NMR and ^{13}C-NMR to dehydrate D-fructose in DMSO solvent to produce 5-HMF, and detected the existence of a key intermediate in Path 1.

From the above discussion, monosaccharides such as fructose and glucose are the direct precursors for the preparation of 5-HMF. However, the content of this series of substances in plant solid waste is often very low. Plant solid waste is more challenging than monosaccharide carbohydrates such as fructose and glucose to produce 5-HMF as a raw material due to its high polymerization and difficult decomposition characteristics. Therefore, from the perspective of raw material selection, plant solid wastes with high content of monosaccharides, cellulose, and hemicellulose have higher 5-HMF conversion potential. Without adding a catalyst, a raw material with a high proportion of hemicellulose is easier to prepare 5-HMF[5] [6] .

The research using Lewis acid (L acid) as a catalyst is the most outstanding research result of Zhao et al. [7]. In 2007, Zhao et al. published their research results on Science. They found that Lewis acid in room temperature ionic liquid 1-ethyl-3-methylimidazole chloride can effectively catalyze the conversion of glucose to 5-HMF, and the reaction is complete. The yield of the latter 5-HMF is as high as 70%.

As mentioned earlier, sugarcane pith is one of the few plant fiber materials with extremely high hemicellulose content. Sugarcane pith is a waste from the paper industry. Use this renewable resource reasonably, such as using it to prepare 5-HMF. The remaining residue contains a large amount of cellulose and can also be hydrolyzed to prepare glucose. It is of great practical significance and application value to increase the added value of sugarcane, turn waste into treasure, and reduce pollution, and it is urgent to discuss in depth. As described herein, the method of preparing 5-HMF from sugarcane pith using cheap Lewis acid aluminum hydroxide as a catalyst, and then continuing to use the residue, and then preparing glucose from it has not been reported in the relevant literature.

2 Experiment

2.1. Use Jupiter-B Microwave Digester for Microwave Reaction

The maximum microwave power and heating rate of Jupiter-B microwave digester are 800 W and 60°C/min, respectively. In the experiment, 1.000 g of sugarcane pith sample pretreated by grinding and sieving and 10 ml of dilute aluminum hydroxide solution were added to each microwave tube, and then placed in the microwave instrument to carry out the conversion reaction of Lewis acid from glucose to fructose And the subsequent dehydration reaction of fructose.

After the reaction, the microwave tube was taken out of the microwave instrument and immediately placed in an ice-water bath to cool to terminate the reaction. After cooling, a Buchner funnel with quantitative filter paper is used to separate the microwave product from solid-liquid. The solid-phase product obtained from the separation is dried in an oven at 100°C to remove moisture, and sealed and stored for testing; liquid-phase products are temporarily stored for testing. The entire experimental process is shown in Figure 2:

https://doi.org/10.1051/matecconf/202135301011

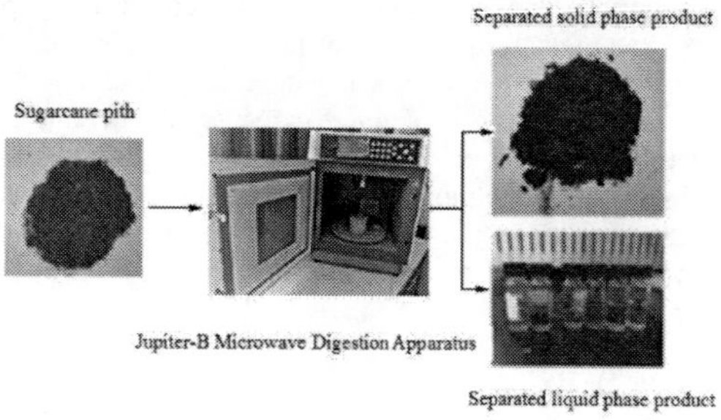

Figure 2. Sugarcane pith microwave-catalyzed conversion of glucose to fructose

The yield of 5-HMF: $Y=100 \times CV/M$. Y is the yield of 5-HMF (%), C is the HMF concentration (mg/L) detected by HPLC, V is the volume of the liquid product after constant volume (L), and M is the dry base of sugarcane pith (mg)

2.2. Use Oxalic Acid to Hydrolyze the Residue for Producing Glucose

Wash the sugarcane pulp residue, dry it, measure an appropriate amount of residue, add 75% oxalic acid in a double amount, and infiltrate it for 30 minutes. Then add about 1000ml of water, place it in a water bath to heat and boil for 1 hour, stirring occasionally. After heating, the product is suction filtered, and the filter residue is discarded. The filtrate is neutralized with calcium carbonate, the generated gypsum is filtered out, the filtrate (ie the sugar solution) is evaporated and concentrated, and then decolorized with activated carbon, and then the sugar solution is concentrated by vacuum distillation. Finally, the sugar solution is cleaned for three days, and the glucose is large. After crystallizing, drying, crushing, and weighing, the glucose is actually obtained, and the final glucose output is the ratio of the amount of glucose weighed after crushing to the residual dregs.

3 Results and discussion

In this study, a chromatographic scan was performed on the liquid product (Figure 3) which obtained from EXPERIMENT 1, and the valuable chemical 5-HMF was found through the comparison of standard materials. After further quantitative analysis, the concentration was 0.9%.

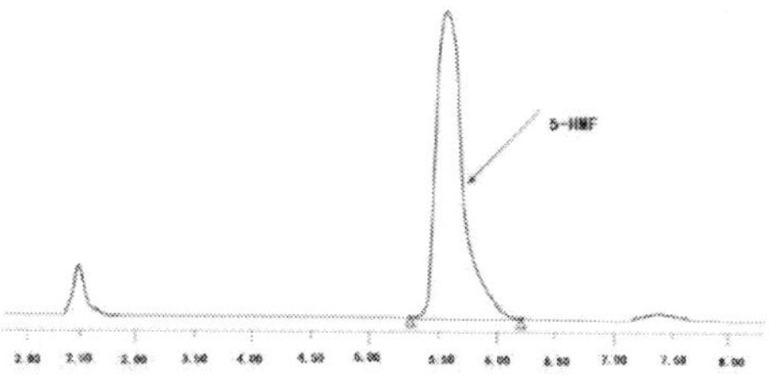

Figure 3. High performance liquid chromatogram of 5-HMF

5-HMF is composed of furan ring, hydroxymethyl group and aldehyde group. It is listed as one of the world's 12 most valuable chemicals. It is an important fine chemical raw material with a pure product value of 10,000 yuan/kg (October 2019 data). The microwave hydrothermal liquid phase products of sugarcane pith are worthy of attention. The main properties of the prepared glucose powder which obtained from EXPERIMENT 2 was detected. Its color: white; dissolution: soluble in a small amount of water, clear and transparent; moisture: 9.10%; glucose: 86.7%; ash: 2.01; dextrin and other aldehydes 2.19%. The ash is mainly gypsum, which has a slightly higher content, but gypsum can be completely removed by ion exchangers after the gypsum is filtered out during the concentration process.

4 Conclusion

Although the experiment obtained high purity 5-HMF, the high-value liquid phase product 5-HMF obtained in this experiment has a low yield. Related research has found that the formation process of 5-HMF is affected by many factors. It can be seen that 5-HMF, a microwave hydrothermal liquid phase product of sugarcane pith, has huge room for improvement.

The quality of sugarcane pulp residue is very good. It is saccharified with 75% oxalic acid to easily produce white glucose powder. The glucose yield is 53.3% to the residue. It is expected that improved operating conditions can also increase the yield.

Acknowledgements

The authors gratefully appreciate the financial support of the High-level Talent Research Startup Fund of Yulin Normal University through contract No. G2018024.

References

1. Y. R. Leshkov, I. N. Chheda and J.A. Dumesic, Science 312(30): 1933-1937 (2006).
2. X. L. Tong, Y. Ma and Y. D. Li, Appl. Catal. A: Gen. 385 (1/ 2), 1-13 (2010).
3. A. Mukherjee, M. J. Dumont and V. Raghavan, BIOMASS BIOENERG. 72, 143-183 (2015).
4. R. Amar, A. Aseka, L. Williamsl, C . Ebede, Carbohydr. Res. 343(18): 3021-3024(2008).
5. B. K. Ozsel, D. Ozturk and B. Nis, Bioresour. Technol. 289: 121627-121633(2019).
6. Z. Chi, X. Zhao and T. Daugaard, BIOMASS BIOENERG. 120, 107-116 (2019).
7. H. Zhao, J. E. Holladay and H. Brown, Science 316 (15): 1597-1600 (2007).

Analysis of 120m Circular Enclosed Coal Yard Based on Finite Element Mechanics

*Wang Teng**

State Nuclear Electric Power Planning, Design & Research Institute CO., LTD., Beijing, China

Abstract. Large diameter circular closed coal yard has large coal pile, high coal pile height, and the cancellation of expansion joints, which has many adverse effects on the structure. The study of finite element analysis method, in view of the large stack on the effect of pile and pile of coal load and temperature effect on structure to do the analysis, the influence of material parameters was verified the correctness of the structure internal force under the condition of various loads is obtained as a result, the temperature effect may produce several times the pile of coal load internal force, provides reference for engineering design.

Key words: closed coal yard; Temperature action; The finite element

1 Introduction

Large diameter circular closed coal yard has the characteristics of small occupation, favorable to environmental protection, flexible operation and so on, and has been applied more and more. The early round coal yard usually set expansion joints along the retaining wall, using reinforced concrete retaining wall and reinforced concrete buttress scheme, similar to retaining wall, although the scheme is simple in calculation theory, but the amount of reinforced concrete, high cost. Laizhou power plant round coal yard retaining wall to cancel the expansion joint, make full use of ring tension resistance coal pressure, advanced technology, less floor area, unit area storage is larger, less reinforced concrete consumption, cost saving. However, the arrangement of the expansion joint will make the structure produce a lot of internal forces under the action of temperature, and the current specifications and commonly used engineering design software can not accurately analyze the effect of temperature on the structure.

Circular coal yard is rarely studied abroad, and almost all existing projects are analyzed and designed independently or with the participation of Wuhan University. With the application of circular coal yard, completing independent design has become an urgent requirement. The project uses the general finite element analysis software to analyze the circular coal yard under the coal load and various temperatures to obtain the internal force of the structure and complete the engineering design.

2. Comparison between finite element analysis and pile test results

Using numerical calculation based on the finite element software, especially as it comes to the elastic-plastic nonlinear calculation, the calculation results of real or not largely depends on the selection of calculation parameters correctly or not, therefore, using the finite element simulation test pile process, compared with pile test report again, to determine the selection of calculation parameters correctly or not is very necessary.

(1) Physical model and soil material parameters

The material of pile body is C40 concrete, which adopts linear elastic constitutive model without considering its cracking and plasticity. Ideal elastic-plastic soil layer, D-P yield criterion and associated flow rule are adopted. Table 3 shows the physical parameters of each soil layer. The interaction between pile and soil is simulated by contact attribute, friction coefficient is constant value 0.3, and the pile bottom and soil layer are coupled by node freedom. Large deformation effects are considered. As the extended D-P plastic yield criterion is used in the calculation, the c and values in the original physical parameters of the soil layer need to be transformed, and the transformation formula is as follows:

$$\tan\beta = \frac{6\sin\phi}{3-\sin\phi} \qquad \sigma_c = 2c\frac{\cos\phi}{1-\sin\phi}$$

$$K = \frac{3-\sin\phi}{3+\sin\phi}$$

C Cohesion of materials

Material internal friction Angle

* Corresponding author: wangteng@snpdri.com

The internal friction Angle of the transformed material
Tensile limit of transformed material
K flow stress ratio
(2) Geometric dimensions
Since there is only vertical load and no horizontal pressure in the soil during pile test, the local pile-soil model of 18° is selected for calculation according to the pile foundation construction situation in the coal yard of Laizhou Power Plant, as shown in Figure 1.

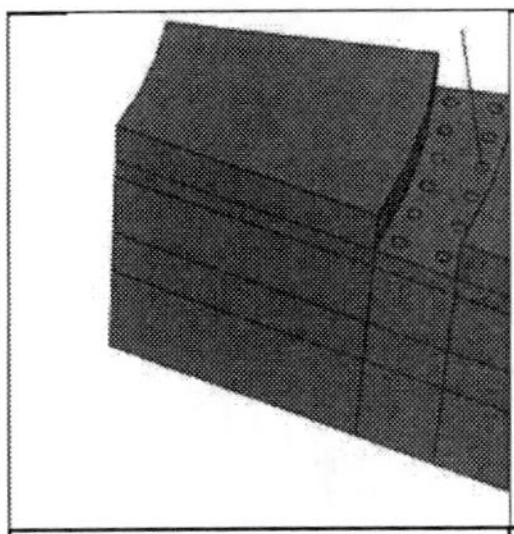

Figure 1 Pile-soil model

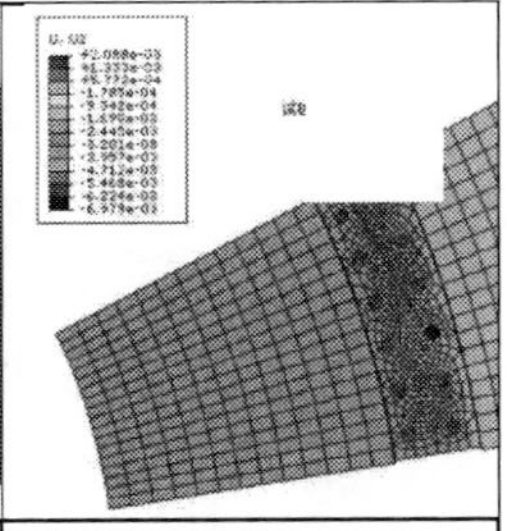

Figure 2 Settlement diagram of test pile

The radial length of the soil is 46.25m, the depth (thickness) is 20m, and the coal yard is divided into four layers both inside and outside the coal yard. In the coal yard: the thickness of crushed stone is 4.4m, the thickness of coarse sand is 5m, the thickness of strongly weathered rock is 3.5m, and the thickness of moderately weathered rock is 7.1m; Outside the coal yard: backfill soil is 4.4m thick, coarse sand layer is 5m thick, strongly weathered rock is 3.5m thick, moderately weathered rock is 7.1m thick.
The pile diameter is 1m, with one row every 3° along the circumferential direction and two rows of piles along the radial direction, with a distance of 3.2m. In this calculation model, there are 12 piles in total, and the middle pile is selected for pile test analysis, as shown in Figure 3.
(3) Load and boundary conditions
In the finite element model, the radial and circumferential displacements were constrained respectively around the soil, and the bottom was fully constrained, that is, the displacements in all three directions were constrained. The bottom of pile is coupled with the degree of freedom of soil, that is, the degree of freedom is bound, and the pile is in friction contact with the soil.
Firstly, the dead weight of soil is calculated without considering the pile (i.e. killing the pile), and the initial stress field of soil layer is obtained. Then the pile is activated and the test load of single pile is applied.
(4) Calculation results and analysis
According to the Test Report of Bored Pile Engineering in No.2 Round Coal Yard of Huadian International Laizhou Power Plant Phase I (2×1000MW) Unit Project, the maximum test load on a single pile is 7000kN, and the test pile number is 227. The test results are compared with the calculation results. The measured vertical maximum displacement value is 10.61mm, the maximum springback deformation is 5.96mm, and the springback rate is 56.2%. The maximum vertical displacement calculated by finite element method is 6.41mm, the springback deformation is 6.01mm, and the springback rate is 93.8%. FIG. 2-2 shows the comparison of P-S curves between finite element calculation and pile test report.
The comparison between the above finite element simulation test pile calculation results and test pile report results can be obtained as follows:
The maximum settlement of pile obtained by field test is 10.61mm, and the maximum settlement of pile obtained by finite element calculation is 6.41mm, which is 60.4% of the maximum settlement of field test.
The springback deformation obtained by field test is 5.96mm, the residual deformation obtained by finite element calculation is 0.4mm, and the finite element calculation result is 6.7% of the field test result.
It can be seen from the above results that the maximum settlement of pile obtained in field test is 4.2mm larger than the calculated settlement, and the residual deformation is 4.46mm larger than the calculated result, and the residual deformation is mainly caused by the sediment at the bottom of pile, indicating that the sediment at the bottom of pile is 5mm~6mm. Removing the influence of the sediment at the bottom of pile, under the action of 7000kN load, The deformation of the pile in the field test is 5.96mm, while the deformation of the calculated pile is about 6.01mm. It can be seen that the calculation results are in good agreement with the field test results. It shows that the calculation model and material parameters can better reflect the actual engineering situation.

3.Foundation treatment analysis of retaining wall of coal yard

The diameter of the coal yard is 120 meters, and the height is 17 meters. The retaining wall of coal yard adopts the integral ring structure. In order to avoid the adverse effect of uneven settlement on the upper cylinder structure of the retaining wall foundation, bored piles are used in the retaining wall foundation of the whole coal yard, and the moderately weathered granite is used as the pile end bearing layer.
(1) Geometric dimensions
Since the structure of retaining wall has little influence on pile foundation, the size of retaining wall is simplified in calculation. The geometric dimensions of the coal yard are as follows:
Retaining wall: upper width 0.6m, lower width 0.8m, linear change in the middle, the total height of 18.2m;
Cap: width 5.6m, height 1.8m. Retaining walls and piles are shown in Figure 3.
For pile foundation construction in coal yard, a local pile-soil model of 180° is selected for calculation. The radial length of soil is 180m, including 120m inside the retaining wall, 30m on both sides outside the retaining wall, 20m depth (thickness), and both inside and outside the coal yard are divided into four layers. In the coal yard: the thickness of crushed stone is 4.4m, the thickness of coarse sand is 5m, the thickness of strongly weathered rock is 3.5m, and the thickness of moderately weathered rock is

7.1m; Outside the coal yard: backfill soil is 4.4m thick, coarse sand layer is 5m thick, strongly weathered rock is 3.5m thick and moderately weathered rock is 7.1m thick, as shown in Figure 4.

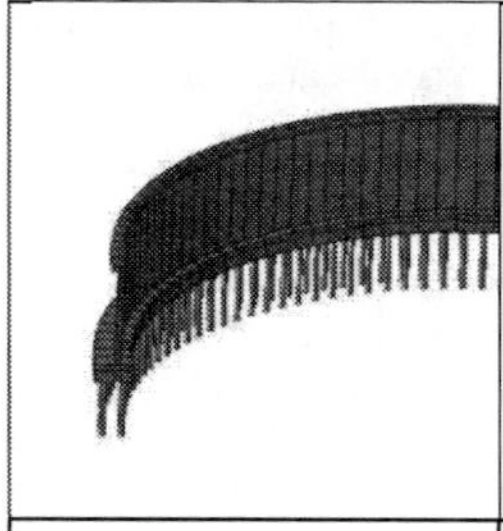

Figure 3 Retaining wall and pile

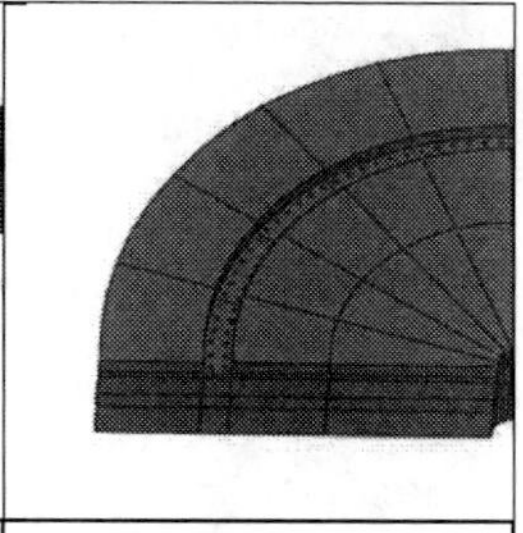

Figure 4 Soil under cap

(2) Stress and deformation of soil

In the result processing, the results show that the local coordinate system, namely the cylindrical coordinate system with the origin of the coordinate in the center of the silo, is adopted: Z direction is vertically downward (direction 3), R direction is radial direction (direction 1), T direction is circumferential direction (direction 2), and the following is the same, which will not be described again. In all figures, the unit of deformation is m and the unit of stress is Pa.

From Figure 5 to Figure 3-8, it can be seen that:

A) Vertical settlement of soil

In the range of 90° of the model coal pile, the soil has settlement, and the maximum settlement occurs at the top of the coal pile, the maximum settlement is 142mm. The settlement of 90° region without coal pile is very small.

B) Radial displacement of soil

The maximum radial displacement of the soil occurs on the surface of the cap bed in the 90° region of the coal pile and is directed to the outside of the coal yard. In this region, the radial displacement of the soil around the pile is small and the maximum horizontal displacement is 28.9mm, indicating that the horizontal extrusion of the soil after dynamic compaction is limited.

Figure 6 is the vertical section of the foundation intercepted where the soil layer's radial displacement is large. It can be seen that the soil layer's radial displacement is large at a depth of 5m near the center of the coal yard.

Since this calculation model is a 180° model with a total of 120 piles, the axial stress distribution of the pile body is different. Therefore, the pile row with the maximum stress on the pile top is selected for analysis in the following result arrangement (as shown in Figure 7). The pile near the coal pile area is called the inner pile and the other one is called the outer pile), which are shown in FIG. 8.

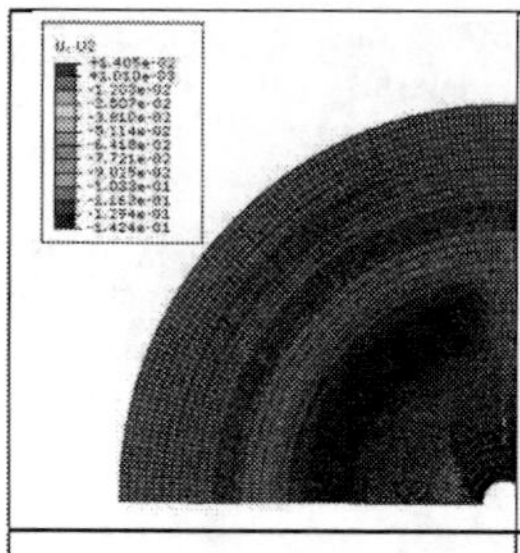

Figure 5 Retaining wall and pile

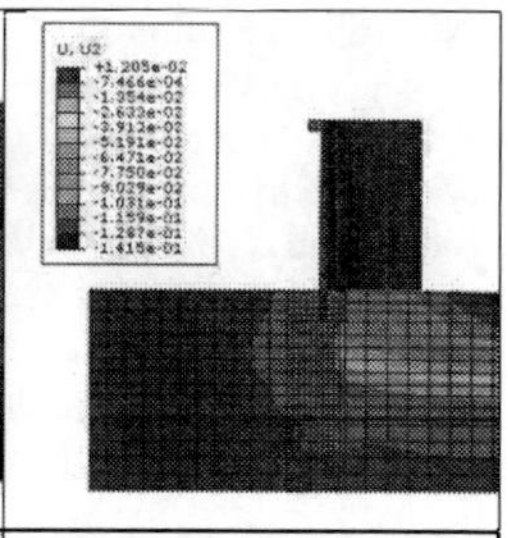

Figure 6 Soil under cap

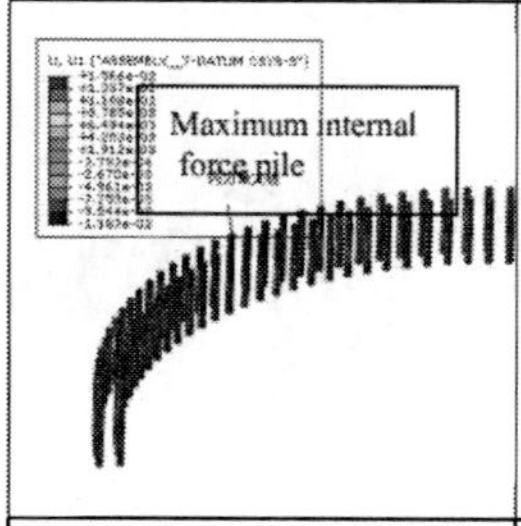

Figure 7 Diagram of pile radial displacement

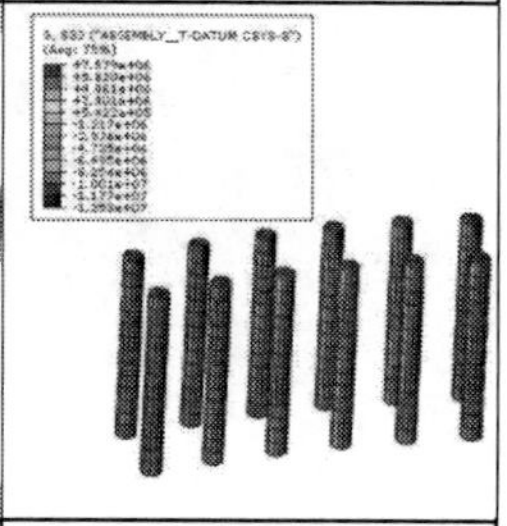

Figure 8 Cloud diagram of pile axial stress

4. Conclusions and Prospects

Based on the analysis of the round coal yard of Laizhou Power Plant, the following conclusions are drawn:

(1) The model and material parameters used in finite element analysis and calculation can better reflect the actual engineering situation.

(2) Soil mass in coal pile area has a great influence on pile stress. In the site with poor geological conditions, foundation treatment is needed to control pile stress. For the site where the bearing layer is not very deep, dynamic compaction method can better control the action of soil on pile in coal pile area.

(3) The effect of environment and service temperature greatly exceeds the requirements of "Code for Design of Reinforced Concrete Silos" (GB50077-2003), and the empirical data provided by the code cannot be used. Because the meteorological conditions of different power plants are different, the effect of temperature must be analyzed separately.

(4) The oxidation of coal pile will produce high temperature. Even if there is no spontaneous combustion, the local temperature inside the retaining wall will increase, and the temperature outside the retaining wall will remain unchanged. Therefore, the insulation material must be made inside the retaining wall to reduce the concrete temperature near the coal pile area.

(5) The amount of reinforcement per unit volume of circular coal yard is much, so attention should be paid to the early stage of the project.

Outlook:

With the enhancement of environmental protection and the reduction of land resources, the application of circular coal yard will be more and more extensive. At present, the main reason for limiting application is first of all, the cost

is relatively high, followed by a relatively long period of time. In the future project, the following items can be considered to reduce the cost:
(1) In cooperation with the technology professionals, the elevation of the coal yard is lowered to the outdoor ground, and the lateral pressure of the soil is used to offset the load of the coal pile. Meanwhile, the seasonal temperature rise and temperature drop can be greatly reduced underground, thus reducing the engineering amount of the whole coal yard.
(2) retaining wall reinforcement is crack control, and the maximum amount of steel in the whole coal yard is retaining wall. The amount of reinforcement can be reduced by using prestressed concrete technology. This approach involves concrete theoretical analysis, design, construction and other aspects, whether the application remains to be studied.

References

1. xia guangzheng, xia dongtao, "finite element analysis of giant coal silo under ambient temperature", space structure, 2006,12(1):62-64.
2. wu dongping, xu yun et al., "stress characteristics and optimization design of a new circular coal bunker structure with ring beam" [J]. Journal of wuhan university (engineering science), 2008,42(2):78-80.
3. peng xueping, "internal force analysis of giant coal silo structure based on different foundation calculation models", journal of wuhan university (engineering science), 2005,38(4):63-67.
4. fan zhenzhong, "some problems in structural design of circular closed coal yard", [J] journal of wuhan university (engineering science), 2007,40(sup)158-161.
5. "Investigation Report of Geotechnical Engineering in Construction Drawing Design Stage of Huadian International Laizhou Power Plant (2×1000MW)".
6. Test Report on Bored pile Construction in No.2 Round Coal Yard of Huadian International Laizhou Power Plant (2 × 1000MW) Unit Engineering. Shandong Base Arrow Engineering Quality Inspection Co., LTD.

Calculation Method for Load-Carrying Capacity of Circular Reinforced Con

Tingwei Wang *

State Nuclear Electric Power Planning, Design & Research Institute CO., LTD., Beijing, China

Abstract. Finite element method and fiber model method were used to calculate the load-carrying capacity of the specimens. Based on the experimental and theoretical analysis, simplified calculation method of the load-carrying capacity for this kind of member is proposed. It indicates that finite element method result is relatively small, fiber model method result accords well with the experimental result. Circular reinforced concrete members covered with steel tube presents both the characteristics of reinforced concrete and concrete filled steel tube member, showing higher load-carrying capacity and greater deformability. The load-carrying capacity of circular reinforced concrete members covered with steel tube can be calculated by the means of the method of reinforced concrete member with confined concrete. The result predicted by the simplified method is in good agreement with the experimental result.

Keywords: Reinforced concrete members covered with steel tube; circular section; load-carrying capacity.

1 Preface

In the 1990s, rock-socketed pile, as a specific pile foundation type, has been widely used in China, and is now more and more widely used in port engineering and bridge engineering. In order to reduce the number of piles, large diameter rock-socketed piles are mostly used. Large diameter rock-socketed piles have the advantages of high bearing capacity, large pile diameter, small settlement and good seismic performance. In the process of pile driving, in order to prevent the steel sleeve from rim and warping, a steel sleeve with a large thickness is generally used, but the thick sleeve wall increases the cost of the project. For example, in the wharf project of Yantian Port in Shenzhen and the wharf project of Huaneng Fuzhou Power Plant, the wall thickness of the steel sleeve of rock-socketed pile can be up to 20mm. In the bearing capacity design of rock-socketed pile, due to the lack of theoretical guidance and calculation formula, the bearing capacity of steel sleeve is generally ignored at present, and only the bearing capacity of reinforced concrete pile is considered. In fact, the steel sleeve and the reinforced concrete in the pile bear the load together, even if a certain part of the steel sleeve is corroded, but the remaining part still has a large bearing capacity. Therefore, if the effect of sleeve is considered in the design, the number of reinforcement in the pile can be reduced and the project cost can be greatly reduced. The member composed of steel sleeve and reinforced concrete in pile is steel tubular reinforced concrete member. There are relevant standards and design specifications for CFST and RC, but there is no calculation method for the bearing capacity of CFST members. At present, there are few researches on the bearing capacity of CFST members. Some existing researches [1-6] mainly focus on the axial compression performance of CFST members. The research on eccentric compression behavior of steel tubular reinforced concrete members is lacking. Based on the experimental study and theoretical analysis of 8 circular section steel tube reinforced concrete (CFST) specimens, a calculation method for the bearing capacity of CFST members is proposed in this paper.

2 Theoretical Analysis

2.1 Theoretical analysis methods

In this paper, three methods are used to calculate the bearing capacity of steel tubular reinforced concrete members. The material constitutive model, strength model and parameters used in the specific method and analysis are as follows.

* Corresponding author: wangtingwei@snpdri.com

2.1.1 Finite element method

The finite element method used Abaqus software, steel and rebar used the isotropic elastic-plastic model provided by Abaqus software, and concrete used the plastic damage model provided by Abaqus software.

2.1.2 Fiber model method

The fiber model method can accurately simulate the process from loading to failure of reinforced concrete members and concrete-filled steel tubular members, and the key is to reasonably determine the constitutive relations of each material.

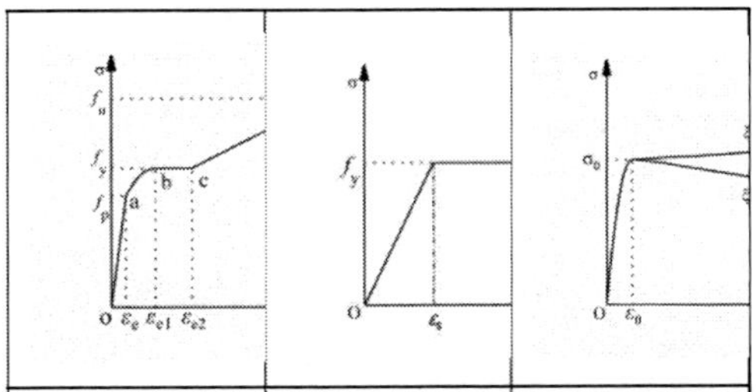

(a) Steel | (b) steel bar | (c) steel tube confined concrete

Fig.1 Constitutive relationship of the materials

The steel adopts the stress-strain relationship expressed in the following formula, as shown in FIG. 1(a) [7] :

$$\sigma_{st} = \begin{cases} E_s\varepsilon_s & (\varepsilon_s \le \varepsilon_e) \\ -A\varepsilon_s^2 + B\varepsilon_s + C & (\varepsilon_e < \varepsilon_s \le \varepsilon_{e1}) \\ f_{yt} & (\varepsilon_{e1} < \varepsilon_s \le \varepsilon_{e2}) \\ f_{yt}\left(1 + 0.6\dfrac{\varepsilon_s - \varepsilon_{e2}}{\varepsilon_{e3} - \varepsilon_{e2}}\right) & (\varepsilon_{e2} < \varepsilon_s \le \varepsilon_{e3}) \\ 1.6 f_{yt} & (\varepsilon_{e3} < \varepsilon_s) \end{cases} \quad (1)$$

In

$$A = 0.2 f_{yt} / (\varepsilon_{e1} - \varepsilon_e)^2,\ B = 2A\varepsilon_{e1},$$
$$C = 0.8 f_{yt} + A\varepsilon_e^2 - B\varepsilon_e$$

Where, σ_s is steel stress; E_s is the elastic modulus of steel, take 206000MPa; ε_s is steel strain; ε_ε is the strain corresponding to the limit of steel proportion, $\varepsilon_e = 0.8 f_y / E_s$. ε_{e1} is the strain corresponding to the yield limit of steel, $\varepsilon_{e1} = 1.5\varepsilon_e$; ε_{e2} is the strain corresponding to the end point of the plastic section of the steel stress-strain curve, $\varepsilon_{e2} = 10\varepsilon_{e1}$; ε_{e3} is the strain corresponding to the ultimate tensile strength of steel, $\varepsilon_{e3} = 100\varepsilon_{e1}$.

The stress-strain relationship of rebar adopts the linear-complete plastic model [8], as shown in FIG. 1(b), namely:

$$\sigma_s = \begin{cases} E_s\varepsilon_s & (0 < \varepsilon_s \le \varepsilon_y) \\ f_y & (\varepsilon_y < \varepsilon_s) \end{cases} \quad (2)$$

Where, ε_s is the reinforcement strain; ε_y is the yield strain of reinforcement, $\varepsilon_y = f_y / E_s$; E_s is the elastic modulus of the steel bar, which is 200000MPa.

The stress-strain relationship model of circular concrete-filled steel tube core concrete in literature [7] is adopted for concrete, as shown in FIG. 5(c), namely:

$$y = 2x - x^2 \quad (x \le 1)$$
$$y = \begin{cases} 1 + q\cdot(x^{0.1\xi} - 1) & (\xi \ge 1.12) \\ \dfrac{x}{\beta(x-1)^2 + x} & (\xi < 1.12) \end{cases} \quad (x > 1) \quad (3)$$

In $$x = \frac{\varepsilon}{\varepsilon_0},\ y = \frac{\sigma}{\sigma_0},\ \xi = \frac{A_{st} f_{yt}}{A_c f_{ck}}$$

Where, σ 、 ε are compressive stress and compressive strain of concrete; σ_0 、 ε_0 is peak compressive stress and peak compressive strain of concrete; ξ is the constraint effect coefficient of CFST; f_{yt} 、 A_{st} is the yield strength and sectional area of steel pipe; f_{ck} 、 A_c are the standard compressive strength and sectional area of concrete.

The fiber model method considers the stress on steel pipe and the constraint effect on concrete separately, which belongs to one-dimensional stress analysis considering transverse constraint. Compared with the finite element method (three-dimensional stress analysis), the calculation is simple.

2.1.3 Simplified calculation method

The above two methods are suitable for the study of load-deformation process and stress mechanism of steel tubular reinforced concrete members, but they are not convenient for engineering design because of the complexity of calculation. Therefore, a simplified bearing capacity calculation method suitable for engineering application should be studied. Although there are special design codes for reinforced concrete and concrete filled steel tube, the methods in these codes cannot be directly used to calculate the bearing capacity of reinforced concrete filled steel tube. Considering that steel casing is usually neglected in the design of rock-socketed piles in the past, the calculation method of the bearing capacity of steel tubular reinforced concrete members is more appropriate. Code for Design of Concrete Structures in Water Transport Engineering (JTS 151-2011[9]) gives the calculation method of the bearing capacity of circular section reinforced concrete members. The formula is simplified on the basis of theoretical analysis, that is, when there are no less than 6 steel bars uniformly distributed along the circumference, the steel bars are continuously converted into steel rings with the same area. For the circular section steel tube reinforced concrete member, different from the ordinary reinforced concrete member, the concrete is constrained by the steel pipe, and the steel pipe also directly participates in the force. Therefore, in theory, steel pipe can be regarded as reinforcing steel, while concrete can be regarded as restrained concrete. In other words, the simplified steel tubular reinforced concrete is considered as double steel bars, and the concrete is calculated according to the strength of constrained concrete . The calculation formula is:

$$N_u = \alpha\alpha_1 f_{cc} A_c \left(1 - \frac{\sin 2\pi\alpha}{2\pi\alpha}\right) + (\alpha - \alpha_t)(f_y A_s + f_{yt} A_{st}) \tag{4}$$

$$M_u = \eta N_u e_i = \frac{2}{3}\alpha_1 f_{cc} A_c r_c \frac{\sin^3 \pi\alpha}{\pi} + (f_y A_s r_s + f_{yt} A_{st} r_{st})\left(\frac{\sin \pi\alpha + \sin \pi\alpha_t}{\pi}\right) \tag{5}$$

$$\alpha_t = 1.25 - 2\alpha,\ e_i = e_0 + e_a$$

Where, N_u 、 M_u are the allowable axial force and bending moment of steel tubular reinforced concrete members; f_{cc} is the strength of steel tube confined concrete, calculated according to the following formula:

$$f_{cc} = \left[1 + (-0.054 \cdot \xi^2 + 0.4 \cdot \xi) \cdot \left(\frac{24}{f_c}\right)^{0.45}\right] f_c \tag{6}$$

f_c is the axial compressive strength of concrete; f_y 、 A_s 、 r_s is the yield strength, sectional area and radius of the circumference of the reinforcement; f_{yt} 、 A_{st} 、 r_{st} are the yield strength, sectional area and radius of the circumference of the steel pipe; e_0 、 e_a 、 η Is the initial eccentricity, additional eccentricity and eccentricity increase coefficient of components, which are determined according to Code for Concrete Structure Design of Water Transport Engineering (JTS 151-2011) [9].

For the case of axial compression, formula (4) is changed by referring to reference [9]

$$N_u = 0.9\varphi\left(f_{cc} A_c + f_y A_s + f_{yt} A_{st}\right) \tag{7}$$

Where, φ is the stability coefficient, and is valued according to literature [9].

3 Conclusion

In this paper, the experimental research and theoretical analysis of the bearing capacity of circular section steel tube reinforced concrete members are carried out. The study concluded as follows:

(1) When calculating the bearing capacity of the circular section steel tubular reinforced concrete member, the circular section steel tubular reinforced concrete member can be equivalent to the double layer reinforced concrete member for calculation.

(2) The double-layer reinforcement constraint can delay the failure of the core concrete and make the steel tubular reinforced concrete members show high deformation capacity.

(3) The simplified calculation method of the bearing capacity of circular section steel tube reinforced concrete members proposed in this paper has simple calculation and high precision, which is suitable for engineering application.

References

1. SHU Gan-ping, LIU Xiao-ying, MIAO Wei. Experimental research and bearing capacity analysis of axial compressive reinforced concrete-filled steel tube short column[J]. Industrial construction, 2010, 40(4): 100-106. (in Chinese)
2. HAN Jin-sheng, DONG Yu-li, XU Zhao-dong et,al. Analysis of axial compression performance for reinforced concrete-filled tubular steel[J].Journal of Civil , Architectural & Environmental Engineering , 2009, 31(3): 11-17.(in Chinese)
3. HOU Yu-ying. Mechanical analysis of reinforced concrete-filled steel tube[J]. Architecture Design Management, 2010, 27(12): 46-48.(in Chinese)
4. LIU Zhao, ZHAO Jun-hai, WANG Juan et,al. Analysis of axial bearing capacity of reinforced concrete-filled circular steel tube short columns[J]. Journal of Architecture and civil Engineering, 2011, 28(4): 92-96.(in Chinese)
5. WEI Hua, WANG Hai-jun. Investigation on strength of concrete-filled circular steel tubular columns[J].Journal of Harbin Institute Technology, 2007, 39(2): 70-73. (in Chinese)
6. WANG Wei, QU Wei-lian. State of the art of elastic-plastic ultimate loads of concrete filled steel tube[J].

J. of HU ST.(Urban Science Edit ion) , 2004, (2): 44-46.(in Chinese)

7. HAN Lin-hai. Structure of concrete filled steel tube Theory and Practice [M]. Beijing: Science Press, 2004.(in Chinese)
8. GB50017-2017, Code for design of steel structures[S]. (in Chinese)
9. JTS 151-2011, Design code for concrete structures of port and waterway engineering[S]. (in Chinese)

Application of NH_3 / CO_2 refrigerant cooling system in a cold storage project

Dao Zhou, Kang Wang *

CCCC-FHDI ENGINEERING CO., LTD., Guangzhou, 510290, China

Abstract. Taking a project's integrated logistics dry port as an example, the advantages and disadvantages of several commonly used refrigerants and their systems are compared and analyzed from safety, environmental protection and energy saving. The technical principles of the NH3 / CO2 refrigerant cooling system are briefly described, and the introduction is also introduced. The refrigeration system selection, equipment configuration and automatic control system scheme of the project refrigerator were provided, which can provide a reference for the selection of the refrigeration scheme of the cold chain project in the future.

Keywords. NH_3, CO_2, refrigeration, cold chain, environmental protection, safety.

1. Preface

With the advancement of science and technology, the development of refrigeration technology, and people's needs for a better life, more and more cold chain projects have emerged at the historic moment. Cold chain logistics has become a necessity of today's life and is closely related to people's lives. Cold chain engineering has become a popular fashion. It refers to all aspects of the production, transportation, storage, and sales of frozen and refrigerated foods until the consumer consumes them. They are always controlled in a low temperature environment that meets the requirements to reduce food losses and ensure A systematic engineering of food quality. The Guilin Suqiao Dry Port Phase-I project (hereinafter referred to as the "Project") described in this article is just such a cold chain project. The design of the refrigeration system of this project follows many factors such as energy efficiency, safety and environmental protection. CO_2 refrigerant cooling system, in order to achieve high-quality use effect and corresponding balance.

2. Project Overview

The total land area of this project is 185413.8 square meters (about 278.1 acres). It is planned to build a comprehensive logistics dry port integrating container yard, refrigerated warehouse, standard warehouse, parking lot and office area. The construction land area of the first phase is 10,816.54 square meters, and the building area is 4,2532.7 square meters.
The construction content of the first phase includes 2 refrigerated warehouses (including refrigerator rooms), 2 general warehouses (including on-site offices), auto repair streets, gates and gate offices, transformer substations, water supply adjustment stations, maintenance workshops and tool materials warehouses, containers Storage yards, container washing sites, garbage collection sites, car parking lots, car parking lots, retaining walls, fences and roads.
The two refrigerated warehouses are single-story warehouses with a building outline of 120.48m x 43m, a building area of 7770.9 square meters, a column top structural elevation of 8m, and an elevated floor of 1.35m. The 1# refrigerated warehouse is a high-temperature refrigerated warehouse (0 °C to 15 °C), and the 2# refrigerated warehouse is a low-temperature warehouse (-25 °C to -22 °C).

3. Comparison of refrigerants

With the rapid development of the economy, many environmental protection issues have arisen. Therefore, the country has long been on the agenda to strongly advocate the harmonious coexistence of man and nature. For the refrigeration industry, the choice of refrigerant is the first consideration. The refrigerants commonly used in low-temperature refrigeration technology in China mainly include NH_3, CO_2 and Freon [1], and their comparisons are shown in Table 1, Figure 1, and Figure 2:

* Corresponding author: wangkang@fhdigz.com

Table 1. Refrigerant comparison table [2]

Refrigerant / Contrast item	NH_3 Two-stage compression refrigeration	NH_3/CO_2 Refrigerant cooling	Freon refrigeration
Environmental Impact Index	ODP=0, GWP=0 Environmentally friendly refrigerant	NH_3:ODP=0, GWP=0 CO_2:ODP=0, GWP=1 Environmentally friendly refrigerant	R404A:ODP=0, GWP=3750 Or R507A:ODP=0,GWP=3800 Environmentally friendly refrigerant
COP energy efficiency value CT=35℃ ET=-45℃	1.39(2.42) Total low and medium cooling capacity in parentheses	1.22	Related to the type of refrigerant
Safety aspect	Toxic, flammable and explosive Unsafe	Safe	Safe
System ammonia charge	Much	Less Can be concentrated in the computer room	None
Compressor	Ammonia compressor	Ammonia compressor only	Freon compressor
Initial investment	Assuming ammonia system is 1	1.05~1.10	Related to specific configuration
Running cost	Lower	High	General

Note: ODP-potential ozone destruction,
GWP-global warming coefficient value,
CT-condensing temperature,
ET-evaporating temperature.

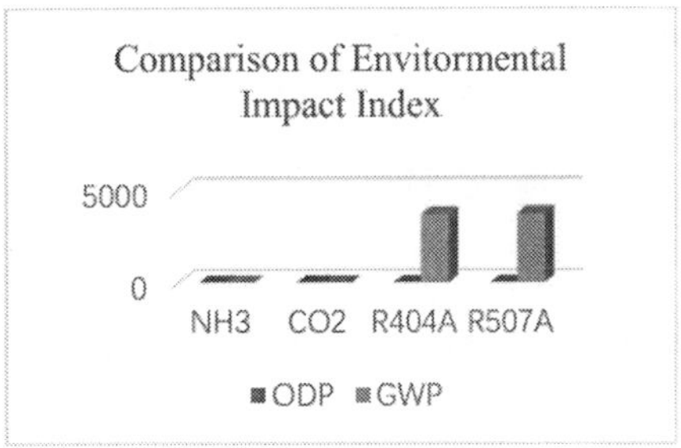

Figure 1. Comparison chart of refrigerant environmental impact index

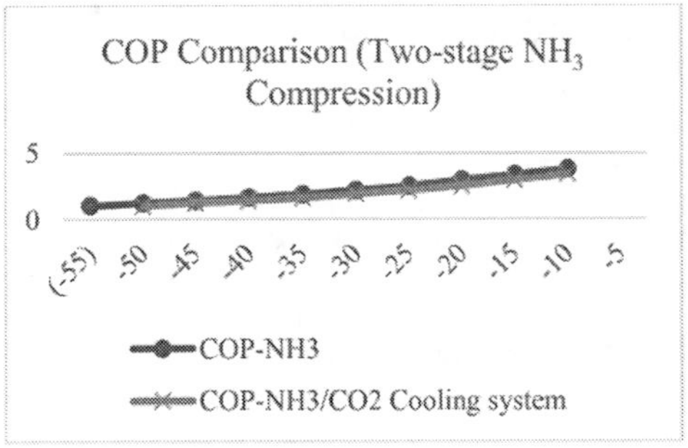

Figure 2. Comparison of COP between two-stage ammonia compression and two-stage ammonia compression CO_2 carrier cooling system [3]

In recent years, the cold storage industry has experienced ammonia leakage accidents in Hebei, Shandong, Jilin, and Shanghai. The government and the public are more concerned about the safety of the refrigeration system than ever before. Chinese national standard "Identification of Major Hazardous Sources of Hazardous Chemicals" GB18218-2018[4] stipulates that the critical amount of ammonia is 10t, that is, when the amount of ammonia in the refrigeration system exceeds 10t, the project is classified as a project with major hazardous sources and all supporting facilities and management measures should be set according to the requirements of major hazard sources. With the NH_3 /CO_2 refrigerant cooling system, the ammonia charge is greatly reduced, and it is concentrated only in the machine room. Its safety is much higher than that of the full ammonia system. At the same time, considering that the Freon listed above is an environmentally friendly refrigerant, its global warming index is still high and will have a certain impact on the environment. From the comprehensive consideration of the social effects of green mountains and green mountains, this project uses NH_3 /CO_2 as the cooling medium of refrigerant cooling system.

4. NH_3 / CO_2 refrigerant refrigeration cycle technology

Schematic diagram of NH_3 / CO_2 refrigerant cooling system is shown in Figure 3 [3]. The system is mainly composed of a refrigeration system and a carrier cooling system. The two systems are independent and both are closed cycle systems. NH_3 refrigeration system consists of ammonia compressor (can be compressed in one stage or two stages, depending on the temperature requirements of the system), oil separator, condenser, ammonia reservoir, throttle valve, ammonia evaporator (condensation evaporator), Ammonia gas-liquid separator and pipeline valves. The CO_2 carrier cooling system consists of a CO_2 reservoir, a CO_2 pump, an evaporator, a CO_2 condenser (condensing evaporator), and a pipeline valve. The condensing evaporator is both the evaporator of the NH_3 refrigeration system and the condenser of the CO_2 carrier cooling system.

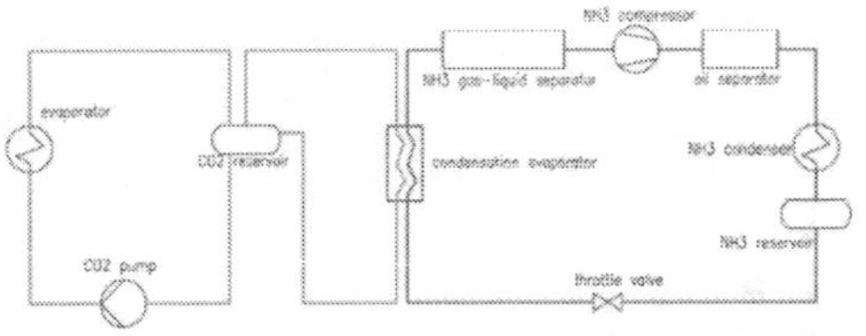

Figure 3. Schematic diagram of NH_3 / CO_2 refrigerant cooling system

Both the high temperature refrigerated warehouse and the low temperature refrigerated warehouse of this project adopt NH_3 / CO_2 refrigerant cooling system. The heat load of the envelope structure, the heat load of the cargo, and the heat load of the cooling fan motor are transmitted to the heat by the evaporator in the warehouse CO_2 refrigerant, the heat absorbed by the CO_2 refrigerant is transferred to the NH_3 refrigeration system through the condensation evaporator, and finally the heat is discharged through the condenser of the NH_3 refrigeration

system to complete the entire heat load transfer process. Because the CO_2 refrigerant needs to be liquefied with NH_3 for heat exchange, two-stage heat exchange temperature difference is required to provide heat exchange capacity. Therefore, the design and optimization of the heat exchanger needs special attention. If the heat exchange temperature difference is large, the system cannot be guaranteed. Higher COP value.
The CO_2 refrigerant system does not need to be equipped with a CO_2 compressor to achieve oil-free operation, which can avoid the reduction of heat exchange efficiency caused by the refrigerated oil entering the evaporator. The CO_2 carrier refrigerant charge in the system is small, and the pipeline size is small. The NH_3 refrigerant is only charged in the equipment room system, and its charge is also small, which can increase the safety of the entire system.

5. Selection of refrigeration equipment

5.1. Freezer Refrigerator

Temperature requirement: -25°C ~ -22°C,
Refrigeration method: NH_3 / CO_2 refrigerant cooling system,
CO_2 refrigerant temperature: -32 °C,
NH_3 evaporation temperature: -35 °C,
NH_3 condensation temperature: + 36 °C,
Cooling equipment form: ceiling type stainless steel tube aluminum fan,
Fan defrost method: automatic water defrosts.
The cooling load is shown in Table 2:

Table 2. Refrigeration load table

No.	Room	Area m^2	Room temperature °C	Cooling equipment load / room kW	Total mechanical load kW
1	1#~ 4#, 6# frozen refrigerated room	495	-25~-22	55	325
2	5#, 7# frozen refrigerated room	742.5	-25~-22	83	

5.2. Cold storage room

Temperature requirement: 0°C ~ 15°C,
Refrigeration method: NH_3 / CO_2 refrigerant cooling system,
CO_2 refrigerant temperature: -7 °C,
NH_3 evaporation temperature: -10 °C,
NH_3 condensation temperature: + 36 °C,
Cooling equipment form: ceiling type stainless steel tube aluminum fan,
Fan defrost method: automatic water defrosts.
The cooling load is shown in Table 3:

Table 3. Refrigerating load of cooling room

No.	Room	Area m^2	Room temperature °C	Cooling equipment load / room kW	Total mechanical load kW
1	1# ~ 4#, 6#cold storage room	495	0~15	166	850
2	5#,7#high temperature storage room	742.5	0~15	240	

5.3. Selection of refrigeration system equipment

5.3.1. Freezer refrigerated room

Three sets of NH_3 / CO_2 unit modules are selected for the low-temperature warehouse refrigeration system, with a total cooling capacity of 110×3 = 330kW, which meets the requirements for use. This unit module includes ammonia compressor, condensing evaporator, high-pressure reservoir and corresponding pipelines, valves, control cabinets, etc.
CO_2 air cooler: ceiling-mounted stainless-steel tube aluminum air cooler is adopted, and automatic water flushing is adopted.

5.3.2. Cold storage room

The high-temperature warehouse refrigeration system uses two sets of NH_3 / CO_2 unit modules, with a total cooling capacity of 428×2 =856 kW, which meets the requirements for use. This unit module includes ammonia compressor, condensing evaporator, high-pressure reservoir and corresponding pipelines, valves, control cabinets, etc.
CO_2 air cooler: ceiling-mounted stainless-steel tube aluminum air cooler is adopted, and automatic water flushing is adopted.

5.4. Frost water system

Four cold air blowers are combined with one defrost pipeline. 3 sets of frosting vertical self-priming pumps are used (2 use and 1 standby), with a single flow of 25m^3 / h and a head of 32m. The frost supply and return water pipes in the refrigerating room are all made of polyurethane insulation, and the outer layer of insulation is made of 0.35mm thick color steel plate. The pipes of the frosting water system are all welded steel pipes.

6. Automatic control system solution

The refrigeration system of this project realizes automatic control. Its control system monitors the relevant refrigerating room temperature, return air temperature, return air pressure, the start and stop of the cold air blower, the alarm of personnel being shut down, etc., and meets

https://doi.org/10.1051/matecconf/202135301014

the various technological requirements of the refrigeration process.

In terms of safety, corresponding gas concentration detection and alarm systems are set up to ensure the safe operation of refrigeration system devices and prevent accidents. Ammonia only circulates in the refrigerating machine room, and an ammonia gas concentration detection alarm system is set in the refrigerating machine room [5]. When the ammonia concentration in the air exceeds the standard, an alarm signal will be automatically issued, and the accident exhaust fan will be automatically activated at the same time. Considering the safety accidents such as CO_2 leakage that may cause suffocation of personnel, CO_2 concentration detectors are installed at locations that may release carbon dioxide, such as CO_2 delivery pumps, valves, flanges, and air-conditioning outlets. Audible and visual alarms are issued when the concentration exceeds the standard. In addition, this project is equipped with multiple portable ammonia concentration detectors and CO_2 concentration detectors, which is convenient for improving the safety of workers.

The control system of this project automatically starts and stops according to changes in external conditions or the user's needs, and automatically adjusts the work of the refrigeration system equipment, which not only saves operators, simplifies management, but also improves the economical operation of the refrigeration equipment. Reduce energy consumption. At the same time, the refrigerator realizes automatic control to keep the temperature of the warehouse stable to improve the quality of commodity storage.

7. Conclusions

NH_3 / CO_2 refrigerant cooling system has its advantages and disadvantages: the two working fluids in the system are environmentally friendly refrigerants, the potential for ozone destruction is 0, and the impact on global warming is minimal. Environmental pollution issues have good economics; the ammonia charge is small, which can reduce the risk of major hazardous sources; the energy efficiency is slightly lower than the ammonia two-stage compression; the critical pressure of CO_2 is high, and the requirements for the equipment are also high.

The comprehensive logistics cold chain project is developing rapidly, and the choice of refrigeration system is very important. It is necessary to choose the appropriate system after weighing various factors in accordance with local laws and regulations and storage production and use requirements.

References

1. Lv Jing, Wang Weifeng, Zhou Chuanyu. Analysis of thermodynamic performance in CO2 cascade refrigeration cycle[J]. Cryogenics,2008,164(4) 38-41.
2. Shi min, Jia lei. CO2 refrigeration technology[J]. Refrigeration and Air Conditioning,2007,12(6) 1-5.
3. Tian Yafen, Zhao Zhaorui, Xing Ziwen. Study on Application Scope for CO2 Cascade Refrigeration System and CO2 Secondary Refrigerant System[J]. Journal of Refrigeration,2016,37(2)22-29.
4. China National Standardization Management Committee. Identification of major hazard installations for hazardous chemicals: GB 18218-2018[S].
5. Lu yaoqing. Practical heating and air conditioning design hand book [M]. (Second Edition). Beijing, China Construction Industry Press,2008.

Particle Behavior Investigation for Different Structures of Hoppers

Panpan Zhang [1, 2], Qingwen Chen [1,*]

[1]China Waterborne Transport Research Institute, Beijing,100088, China.
[2]Key Laboratory of Logistics Equipment & Control Engineering, Beijing 100088, China.

Abstract. In this paper, two different kinds of hoppers were proposed. In order to investigate particle behavior in hoppers, the discrete element method (DEM) was introduced and soybeans were chosen as test material Particle behavior of four different samples were simulated under different flow rate. Results reveal that the deflection cone greatly affected the particle behavior. After adding the deflection cone in the hopper, the trajectory of particle material can be restrained and the dust diffusion can be reduced.different distances of the deflection cone showed different guiding effect on the particle movement.

1 introduction

With the development of new corona-virus, global food trade is extremely active. Shipping is the most important means of international food transportation. When the ship arrives at the port, grain loading and unloading process is also an important area of port environmental protection. The ship is unloaded and grain will enter the silo. if grain is going be distributed, it is necessary to use hoppers for truck loading. The hopper in the distribution station is an important equipment. the performance of the hopper has a great influence on the bulk cargo handling. In this process, it is the prone point of dust shown in Figure 1. In recent years, researchers and some companies focus on the study of particle behavior and dust suppression [1-3].

Fig. 1 The dust caused in the truck loading

The discrete element method known as DEM is now widely applied to investigate the particle behavior in areas of food, mineral and chemical industries.The software packages EDEM were developed based on the discrete element method which can model various materials such as large rocks, fine powders, grains, fibers, and tablets and can provide realistic material behavior to facilitate the equipment development as shown in Figure 2 [4-5].

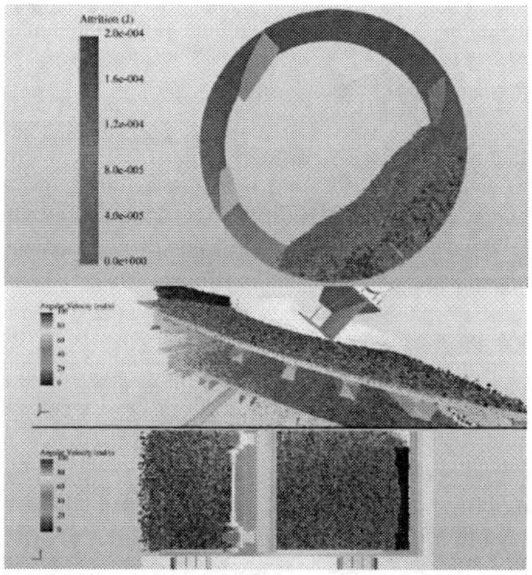

Fig. 2. The application of the software EDEM [6].

In this paper, different structures of hoppers were proposed. the discrete element method was used via the the software package EDEM and investigation was carried out to evaluate the particle behaviors. the influence factors of different deflection cone positions and different material flow rate were simulated.

2 Hopper Design

In order to access the particle behavior of different hoppers, two different kinds of hoppers were designed.

* Corresponding author: chenqw@wti.ac.cn

the hopper shown in Figure 3 is the same as the one used in ports. Unlike the common hopper, a deflection cone as shown in Figure 4 was set in the hopper to change the flow of materials.

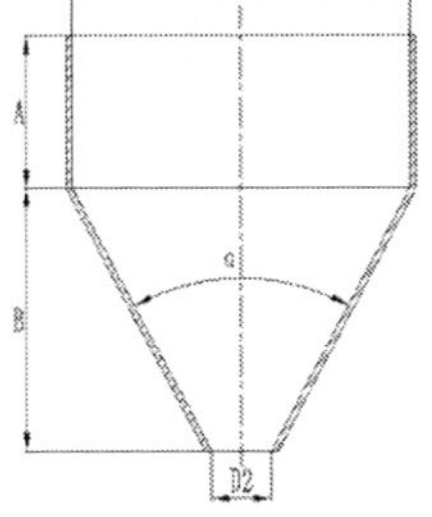

Fig. 3. The structure of the hopper

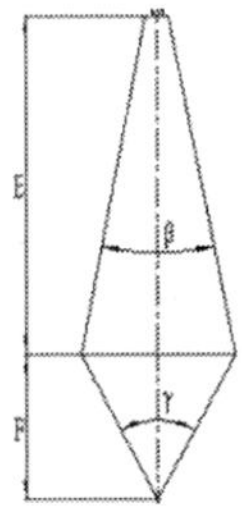

Fig. 4. The structure of the deflection cone

In Figure 3, the diameter of the inlet marked as D1 and the diameter of the outlet marked D2 were set as the fixed value. Paramenters of the hopper were listed in Table 1. The parameterS of the deflection cone were shown in Table 2.

Table 1. Parameters of hoppers

D1 (mm)	D2 (mm)	A (°)	A (mm)	B (mm)
280	50	56	125	215

Table 2. Parameters of a deflection cone

E (mm)	F (mm)	β (°)	γ (°)
280	120	22	56

In order to discover the position influence of the deflection cone in Figure 4 , different heights in were set as shown in Table 3.

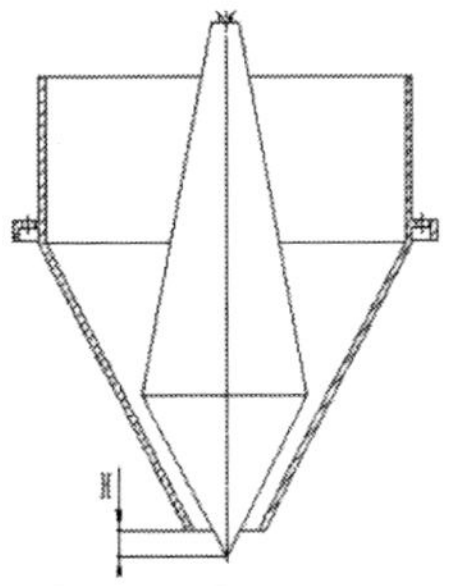

Fig. 4. The structure of the deflection cone

Table 3. Height setting of a deflection cone

Sample 1	Sample 2	Sample 3	Sample 4
No deflection cone	H1 (mm)	H2 (mm)	H3 (mm)
	0	20	30

3 Simulation of different hoppers

The physical parameters of the soybeans and the hopper were shown in Table 4 and Table 5. The equivalent particle diameter of soybeans is random distributed from 3.2mm to 4mm in diameter. During the soybeans unloading simulation process, the mass flow rate of soybeans is 2500 kg/h.

Table 4. Physical parameters of soybeans and the hopper

Item	Density/kg·m-3	Shear modulus /Pa	Poisson's ratio
Soybean	1222	2.0e+07	0.4
Steel	7900	1.0e+08	0.25

Table 5. Contact parameters of different materials

Item	Restitution coefficient	Coefficient of static friction	Coefficient of rolling friction
Soybean-Soybean	0.668	0.2	0.02
Soybean-Steel	0.72	0.2	0.0262

Calculation domain of the hopper was set as a cylinder down the outlet of the hopper with 600mm in height and 600mm in diameter as shown in Figure 5. During the simulation, Hertz-Mindlin non-slip model was adopted.

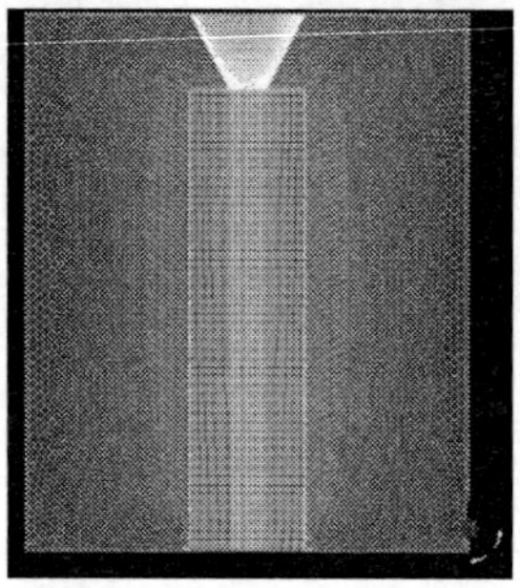

Fig. 5. Calculation domain setting

4 Results and Discussion

Particle behaviors of samples shown in Table 3 were simulated as shown in Figure 6 and Figure 7. the simulation results reveal that the deflection cone can guide the flow of soybeans as well as the dust. However the distance between the deflection cone and the outlet of the hopper have a different and significant influence on the material flow. As the distance increases, the gap between the hopper and the deflection cone is decreasing. Decreasing of the gap can compress the soybean flow which will slow down the flow in a certain range.

From results as mentioned above, it can be seen that the deflection cone can effectively collect the trajectory of soybeans and reduce the dust diffusion. At the low low rate of 1t / h, the particle diffusion range without the deflection cone is the largest ;

At a high material flow rate of 2.5 t / h, the deflection cone has the same effect. Comparing the results of the same deflection cone position under different material flow rates, it can be seen that high material flow rate is also conducive to collecting particle flow and inhibiting dust diffusion. When the material flow rate is high, the collision constraint between particles is more obvious and plays a guiding role. Therefore, even if there is no the deflection cone, the degree of particle diffusion is also significantly weaker than that of low material flow rate.

When the deflection cone is reduced to the position of 30mm, the blocking phenomenon occurs in the hopper, indicating that the annular channel at this position is too narrow to reach the material flow rate of 2.5t / h.

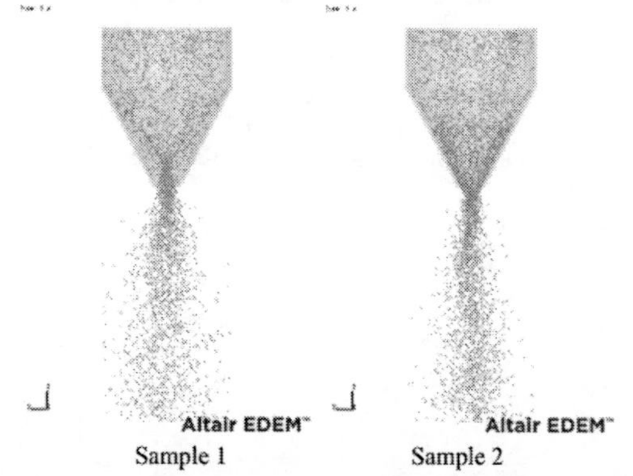

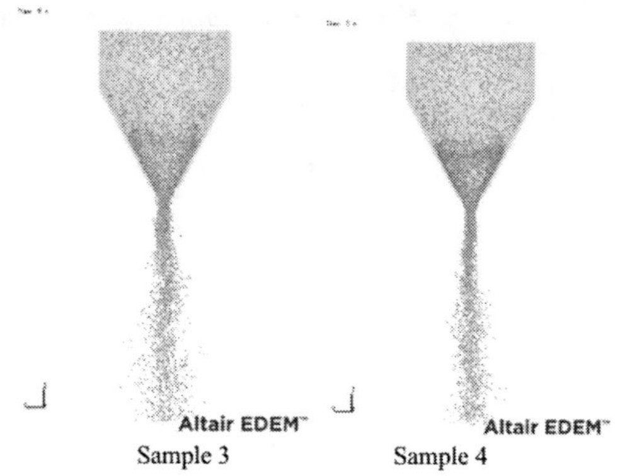

Fig.6. Particle behavior of different samples at the flow of 1.0t/h

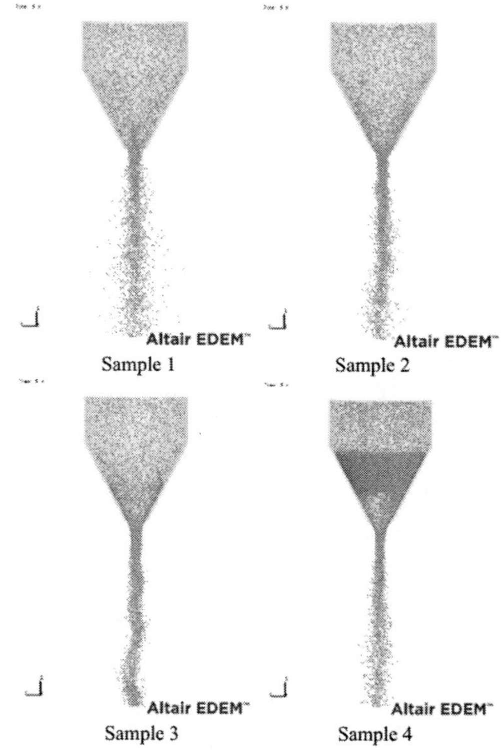

Fig.7. Particle behavior of different samples at the flow of 2.5t/h

Figure 8 demonstrates that after the deflection cone is added, the lower the position, the better the effect of the particle flow. This phenomenon shows that the annular channel formed by the deflection cone and the inner wall of the hopper has a guiding effect on the particle movement, and weakens the severe collision when the particle flow converges, thereby inhibiting the dust diffusion.

https://doi.org/10.1051/matecconf/202135301015

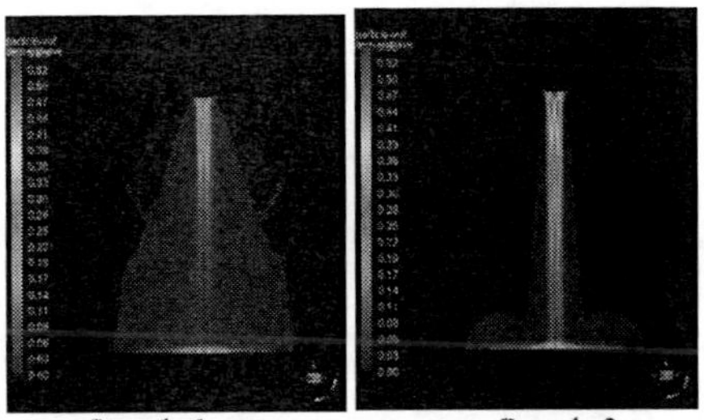

Sample 1 Sample 3

Fig. 8. the percentage of dust particle for different hoppers

5 Summary

Hoppers are important in the bulk cargo distribution. In this paper, different hoppers were proposed a the discrete element method. Results shows that structure of the hopper greatly affected the particle behavior which has an important effect on the performance of hoppers . After adding the deflection cone in the hopper, the trajectory of particle material can be effectively restrained and the dust diffusion can be reduced. The annular channel formed by the deflection cone and the inner wall of the hopper has a guiding effect on the particle movement. In future, more factors that affect performance of hoppers will be studies and provide instruction for hopper design.

References

1. Panpan Zhang, Simulation of Particle Behavior for Different Hoppers, Advances in Engineering Research, Vol.125(2017):228-231.
2. Wypych P., Cook D., Cooper P. Controlling dust emissions and explosion hazards in powder handling plants [J]. Chemical Engineering and processing.2005, 44:323-326.
3. Liu Z.Q., Ma L.Y., Zhao X.Y. . Experimental Study the affection of the Dust Emission on Sand Stockpile by the Height of Windbreak Wall and Wind Speed. Applied Mechanics and Materials, 2014, 823:989-994.
4. Ai, Jun, Jian-Fei Chen, J. Michael Rotter, and Jin Y Ooi. Assessment of rolling resistance models in discrete element simulations. Powder Technology 206 (2011): 269-282.
5. Baran, Oleh, Alfred DeGennaro, Enrique Rame, and Allen Wilkinson. DEM Simulation of a Schulze Ring Shear Tester. Proceedings of the 6th International Conference on Micromechanics of Granular Media. 2009. 409-412.
6. Information on https://www.altair.com/edem/

Preparation of Fiber Reinforced Silica Aerogel Block

Shuning Zhai [1], Chi Cao [1], Yang Liu [1], Chen Gui [1], Kai Yuan [1], Qun Gao [1,], Benjamin Wilson [2]*

[1] School of Shanghai Institute of Technology University, Shanghai, China
[2] School of Nanyang Technological University, Singapore

Abstract. Silica aerogels are lightweight, highly porous nanomaterial with extremely large internal surface area but fragile and brittleness. To enhance the applicability of aerogels, fiber reinforced silica aerogel insulation blocks are prepared by frothing method. Polyurethane emulsion acts as binder whereas polyethylene (PE) fiber and polyurethane (PU) fiber is used as reinforcing phase. The effect of polyethylene fiber and polyurethane fiber on the performance of composites were characterized by material's chemical microscopic morphology, compression performance, recovery performance, surface area, thermal conductivity and hydrophobicity. The research results show that the prepared PU fiber reinforced silica aerogel insulation block has excellent comprehensive performance. The water absorption rate of decreased from 19.35% (0 phr, PU fiber) to 6% (20 phr, PU fiber), and comprehensive rebound rate was increased by 97.83% at 20 phr PU fiber compared with pure insulation block (83.88%). Meanwhile, the dimensional stability of the composite aerogel block is above 96%. PU fiber reinforced silica aerogel insulation block has light weight, good hydrophobicity and good thermal and mechanical properties, so it has a wide range of application prospects in the field of insulation.

1 Introduction

Aerogel has porous structure formed by interconnecting three-dimensional nano network frameworks, so it has high porosity (80% ~ 99.8%)[1], ultra-high specific surface area (800 ~ 1000 m2/g)[2], super low thermal conductivity [0.001~0.043 W/(m·K)][3] and very low density (0.003~0.5 g/cm3)[4]. And it is one of the key materials in the field of high-performance thermal insulation at present. However, pure silica aerogel has low mechanical properties and is brittle, which limits its industrial application. Therefore, it is necessary to strengthen the silica aerogel to prepare the silica aerogel thermal insulation composite material with strong mechanical properties[4].

Kim et al. prepared a glass fiber/silica aerogel composite insulation material and found that glass fiber can effectively enhance the toughness of Silica aerogel[5]. Zhang et al. explored the use of aluminum silicate fiber and glass fiber as silica aerogel reinforcements, and found that the composite filled with fibers and silica aerogel increased in compression strength than the fiber mat alone after 30% compression deformation. The composite material composed of glass fiber and Silica aerogel has more than 200% higher mechanical strength than aluminum silicate fiber[6]. Li et al. compounded sepiolite fibers with silica aerogel, which has good interfacial interaction to improve the mechanical properties of aerogel materials [7]. Zhou et al. added polyester fiber to silica aerogel, and the pressure carried by the material increased by 40 kPa compared with pure silica aerogel when the elastic deformation became 80%[8]. However, silica aerogels are generally prepared into powders through the sol-gel method, solvent replacement and supercritical drying processes, and it is difficult to make blocks and directly use them as thermal insulation materials. Generally, silica aerogel is compounded on the fiber mat to prepare a block with strong mechanical properties, or the aerogel is first made into powder, and then placed between two plates to form a sandwich-like insulation block, the thermal insulation performance of this sandwich product is difficult to reach the ideal level of super thermal insulation materials.

Water-based polyurethane (WPU) not only has the excellent properties of polyurethane (flexibility, adhesion, etc.), but also reduces the amount of organic solvents (or even to zero). It is a new type of environmentally friendly material, so it has received many widespread concern. Polyurethane binders are based on the unique chemical properties of isocyanates. Isocyanate is a compound containing an isocyanate group (-NCO) in the molecule. This group has a highly unsaturated bond structure with overlapping double bonds and can react with various compounds containing active hydrogen.

In this paper, combining silica aerogel with low thermal conductivity and water-based polyurethane emulsion with high cohesiveness, the two are combined by a wet process to prepare silica aerogel insulation block; further with PE Fiber and PU fiber are used as reinforcement, and a fiber-reinforced silica aerogel insulation block with a size of 150 mm × 150 mm × 10 mm is prepared. Also, we compare and study the influence of different fibers content on the structure and performance of materials, in order to obtain thermal insulation blocks with excellent

* Corresponding author: wityy@sit.edu.cn

thermal insulation performance, so that large-scale and environmentally-friendly production of thermal insulation materials becomes possible.

2 Experiment

2.1 Main materials

The raw materials used in this experiment mainly include water-based polyurethane emulsion (Zhongke Huayu Technology Development Company Limited.), sodium dodecyl sulfonate (Titan Technology Company Limited.), Silica aerogel (Shenzhen Zhongning Technology Company Limited.), polyethylene fiber, polypropylene fiber and distilled water.

2.2 Material preparation

Weigh 20 phr of water-based polyurethane emulsion, 2 phr of sodium dodecyl sulfonate, polyethylene fiber and polyurethane fiber and add 0 phr, 5 phr, 10 phr, 15 phr, and 20 phr, respectively. Put the above raw materials into a high-speed mixer and stir for 30 minutes, then add 15 phr of Silica aerogel, and stir again for 30 minutes. The mixed homogeneous slurry is obtained, poured into a mold, and placed in a thermostatic oven for 48 hours at a constant temperature of 50° C. to obtain a fiber-reinforced silica aerogel insulation block.

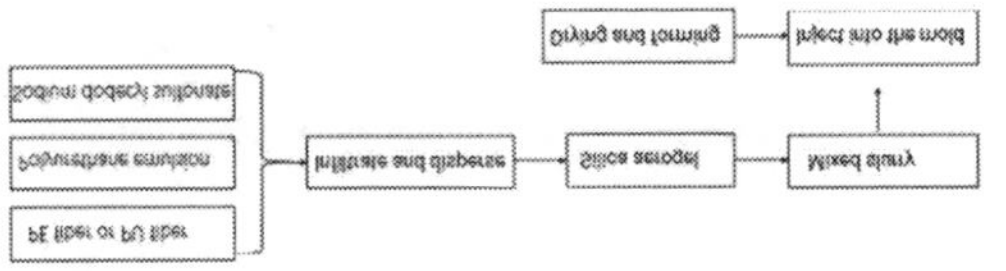

Figure 1. Preparation process

2.3 Testing and characterization

Field emission scanning electron microscope (FESEM) was performed on a S-3400N instrument to observe the microscopic appearance of the sample; the thermal conductivity meter was determined on a DRPL-III instrument to analyze the thermal conductivity of the sample; the water absorption per unit volume of the sample was calculated by water immersion method: 1) measure the mass before water (m1) and after water (m2); 2) water absorption w=[(m2-m1)/m1]*100%; the compressive strength of the sample was measured on a Sun 500 universal material testing machine; the compression resilience rate of the sample was performed on a Sun 500 universal material testing machine: 1) measure the initial thickness h0 and recovery thickness h1; 2) Compression resilience rate R=(h1/h0)*100%; the dimensional stability of the sample was determined by the constant temperature drying method: 1) Use a vernier caliper to measure the size before drying (a0) and the size after drying (a1); 2) Dimensional stability D=(a1/a0)*100% ; Brunner–Emmet–Teller (BET) was performed on a ASAP2460 instrument to observe the surface area of the sample

3 Results and discussion

3.1 The influence of fiber types on the microscopic morphology of silica aerogel insulation blocks

The microscopic morphology of the composite material can intuitively describe the composite situation of fiber and silica aerogel. Figure 2(a) is the SEM image of silica aerogel insulation block prepared without fiber. It can be seen from the figure that a large amount of Silica aerogel is combined through PU emulsion to form needle-shaped particles to form a network structure, and there are a lot of pores inside. It can be seen from figure 2(b) and (c) that PE fiber and PU fiber are embedded in silica aerogel insulation block, while silica aerogel is attached to the surface of two kinds of fibers, which proves that silica aerogel and fiber form a better interface bonding. The interfacial combination provides strong evidence for the improvement of mechanical properties of composite materials. The gap between the fiber and the fiber is large, which reduces the thermal bridge effect during contact and reduces the thermal conductivity of the composite material.

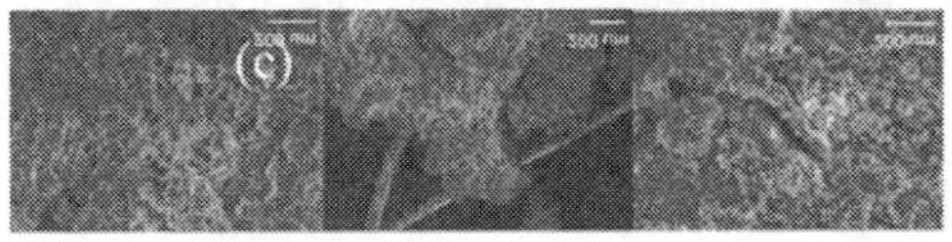

Figure 2. SEM images of silica aerogel insulation blocks: (a) No fiber added; (b) PE fiber (5 phr); (c) PU fiber (5 phr).

3.2 The influence of fiber type on the thermal conductivity of silica aerogel insulation blocks

Thermal conductivity refers to the amount of heat transferred through a unit horizontal cross-sectional area in a unit time when the temperature vertical downward gradient is 1 °C/m. Figure 3 shows the thermal conductivity of the silica aerogel insulation block with different fiber additions. With the increase of PE fiber addition, the thermal conductivity of the composite material gradually increased, reaching a maximum of 0.02953 W/(m·K); however, with the increase of PU fiber, the thermal conductivity of silica aerogel insulation block increases first, then decreases and then increases, reaching a maximum of 0.02387 W/(m·K). The fiber can shorten the solid phase transmission path, reduce the heat resistance, increase the solid phase heat conduction, and increase the thermal conductivity. Therefore, the thermal conductivity of the aerogel insulation block with the fiber is greater than that when it is not added.

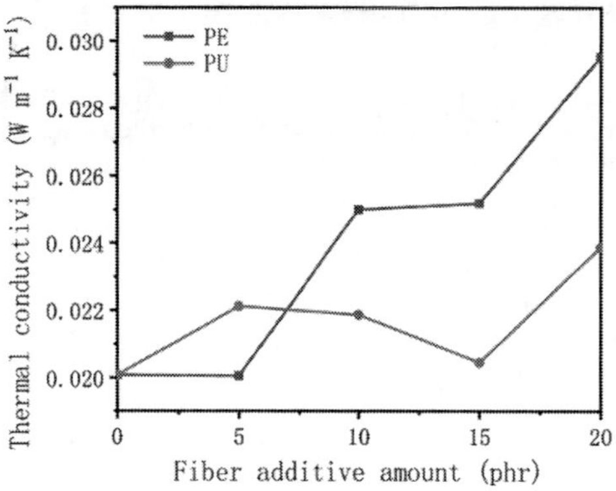

Figure 3. Thermal conductivity image of silica aerogel insulation blocks.

3.3 The influence of fiber type on the water absorption of silica aerogel insulation blocks

Water absorption is a physical quantity that indicates the degree of water absorption by an object under ambient pressure. Figure 4 is a graph of water absorption of silica aerogel insulation blocks with different fiber additions. Polyurethane has weak water absorption as binding agent, and the two fibers have better moisture resistance, the water absorption rate of the silica aerogel insulation block decreases with the fiber content increases. When the amount of PE fiber added is 20 phr, the composite material has a loose structure and more pores, so a large amount of water can be stored inside, resulting in an increase in the water absorption rate of the material. When the amount of PU fiber added is 20 phr, the water absorption of the silica aerogel insulation block reaches the lowest value of 6%.

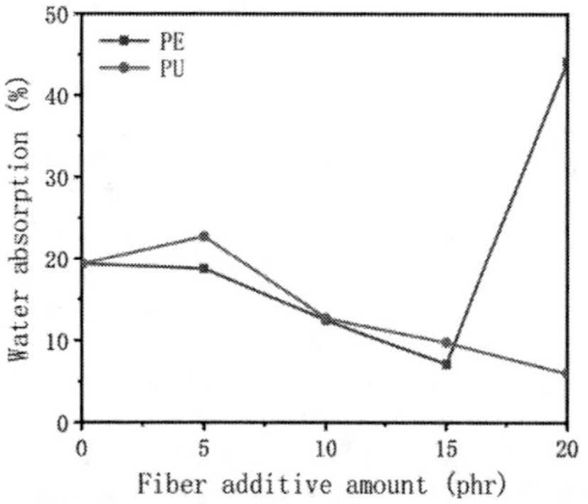

Figure 4. Water absorption image of silica aerogel insulation blocks.

3.4 The influence of fiber types on the compressive strength of silica aerogel insulation blocks

Compressive strength is the compressive stress of the sample when it is deformed by 20%. It characterizes the material's ability to resist compressive loads without failure. Figure 5 is a graph of the compressive strength of silica aerogel insulation blocks with different fiber additions. For PE fiber, the contact area and interface binding force between aerogel and fiber are small, so as the external stress increases, the compressive strength decreases. For PU fiber, when the fiber content is small, the binding force of silica aerogel and fiber is stronger, and its compressive strength increases. When the amount of PU fiber added is 10 phr, the compressive strength of the composite material reaches the maximum value of 0.0323 MPa; when the fiber content is high, the interface between the silica aerogel and the fiber is weakly bonded, and its compressive strength is low. When the PU fiber addition amount is 20 phr, the composite material's compressive strength reaches the minimum value of 0.00575 MPa.

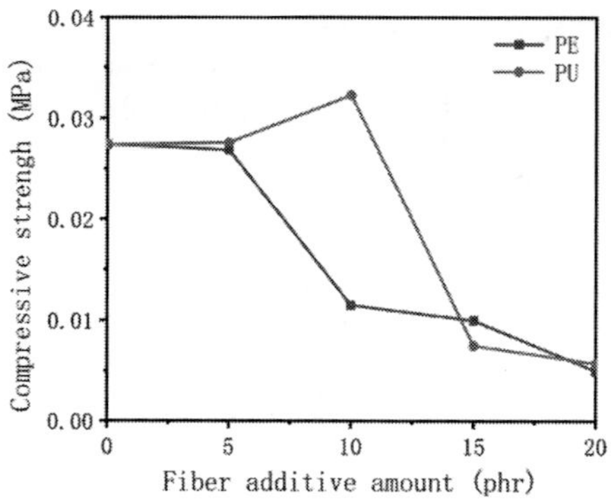

Figure 5. Compressive strength image of silica aerogel insulation blocks.

3.5 The influence of fiber type on the compression rebound of silica aerogel insulation blocks

Compression rebound rate is the degree to which the sample returns to its original state after being deformed by the applied force. Figure 6 shows the compression rebound of the silica aerogel insulation block with different fiber additions. As the amount of fiber increases, the compression rebound of the sample is improved. This is because the silica aerogel is supported by fibers. As the fiber content increases, the static air layer increases. From the side, it shows that the addition of fibers helps to improve the compression rebound of the silica aerogel insulation block. When the amount of PU fiber added is 20 phr, the compression rebound rate of the composite material reaches the maximum value of 97.83%.

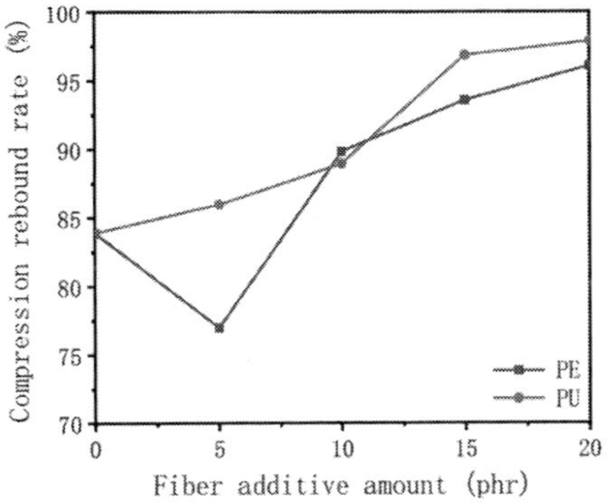

Figure 6. Compression rebound image of silica aerogel insulation blocks.

3.6 The influence of fiber types on the dimensional stability of silica aerogel insulation blocks

Dimensional stability refers to the performance of a material that does not change its external dimensions when it is affected by external conditions, and is an index to determine whether the material is prone to deformation. Figure 7 is a graph showing the dimensional stability of silica aerogel insulation blocks with different fiber additions. It can be found from the figure that the dimensional stability of the two different fibers is above 96%, which proves that the material has good dimensional stability.

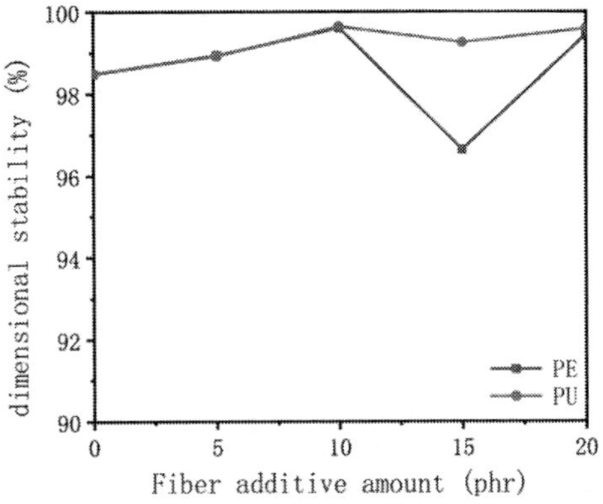

Figure 7. Dimensional stability image of silica aerogel insulation blocks.

3.7 The influence of fiber types on the surface area of silica aerogel insulation blocks

The surface area refers to the total area of a unit mass material. Figure 8 is a graph showing the surface area performance of silica aerogel insulation blocks with different fiber additions. It can be found from the figure that the specific surface area of the aerogel insulation block is the largest (288.32 m2 g-1) when no fiber is added. With the increase of the content of the two fibers, the surface area of the aerogel insulation block is decreasing. The test results show that the addition of fibers has a certain effect on the aerogel insulation block, increasing the pores of the aerogel insulation block and reducing its surface area.

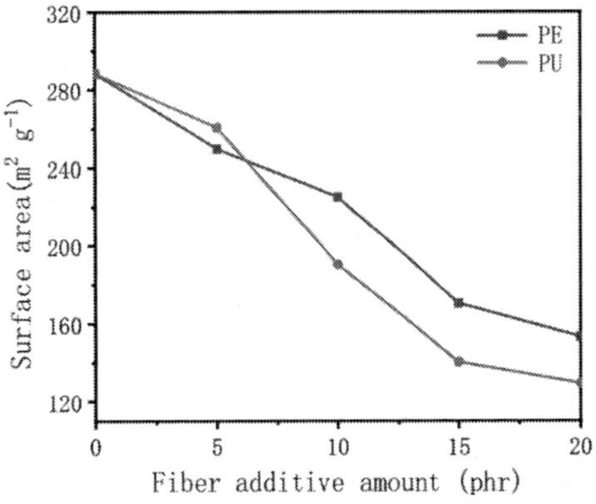

Figure 8. Surface area image of silica aerogel insulation blocks.

4 Conclusion

Using silica aerogel as the powder, polyethylene fiber and polyurethane fiber as the reinforcing phase, and polyurethane emulsion as the binding agent, the silica aerogel insulation blocks with better thermal insulation performance are prepared by the frothing method and ambient drying method.

Explore the effects of polyethylene fiber and polyurethane fiber and their added content on the thermal conductivity, water absorption, compression rebound, surface area, compression strength and dimensional stability of silica aerogel insulation blocks. The results show that the overall performance of the polyurethane fiber reinforced silica aerogel insulation block with 10 phr is the best. The thermal conductivity is 0.02187 W/(m·K), the water absorption rate is 12.73%, the surface area is 190.34 m^2g^{-1}, the compression rebound rate is 81.77%, and the compression strength (20% Strain) is 0.0323 Mpa.

Acknowledgments

This work was financially supported by postgraduate high-level applied scientific research results fund.

References

1. H. Suo, W. Wang, S.J. Jiang, et al, Research progress of high temperature resistant aerogel insulation materials, Shanghai Aerospace. 36 (2019) 61-68.
2. X.Q. Jiang, Research progress of silica aerogels, Fine and Specialty Chemicals. 28 (2020) 42-46.
3. Y.Z. Lu, R. Chen, Y.Z. Li, et al, Hydrophobic modification of silica aerogels and research progress

of composite materials, Shandong Chemical Industry. 49 (2020) 72-73.

4. L. Xu, Y.G. Jiang, J.Z. Feng, et al, Preparation of fiber-reinforced Al2O3- SiO2 aerogel thermal insulation composite material and thermal insulation performance, Journal of Nanjing University of Technology (Natural Science Edition). 42 (2020) 461-466.
5. S. Kim, J. Seo, J. Cha, et al, Chemical retreating for gel-typed aerogel and insulation performance of cement containing aerogel, Construction And Building Materials. 40 (2013) 501-505.
6. Y. Zhang, X.G. Ge, G.J. Lu, et al, Preparation and properties of aluminum silicate fiber and glass fiber composite silica aerogel materials, Inorganic Salt Industry. (2020): 1-5.
7. X. Li, Q. Wang, H. Li, et al, Effect of sepiolite fiber on the structure and properties of the sepiolite/silica aerogel composite, Journal of sol-fel science and technology. 67 (2013) 646- 653.
8. J.L. Zhou, M.J. Shi, D. Long, et al, Preparation and properties of silica aerogel toughened by polyester fiber, Functional Materials. 49 (2018) 10150-10154.

Research review on cutting temperature and cutting vibration of titanium alloy

Yuqing Li [1], Shuncai Li [1,*] and, Qiu Yu [2]

[1] JSNU-SPBPU Institute of Engineering, Jiangsu Normal University, Xuzhou, China
[2] School of Mechanical and Electrical Engineering, Jiangsu Normal University, Xuzhou, China

Abstract. The text should be set to 1 line spacing. The abstract should be centred across the page, indented 17 mm from the left and right page margins and justified. It should not normally exceed 200 words The titanium alloy called "strategic metal" is a typical refractory material that causes high temperature and vibration during cutting due to its high strength, low thermal conductivity and low elastic modulus. In the processing process, the tool wear is serious, and the accompanying cutting heat and cutting vibration are coupled to each other and change with time. Studying the temperature and vibration characteristics in the processing process is conducive to further understanding of cutting mechanism, better controlling temperature and vibration, improving cutting processing quality and reducing production cost. This paper reviews the current situation of cutting temperature and cutting vibration at home and abroad. Scholars focus on the influence of cutting factors on temperature and vibration, based on finite element simulation and mathematical modeling, and the research on temperature and vibration coupling of titanium alloy cutting should be further strengthened.

1 Introduction

Due to its high specific strength, excellent chemical corrosion resistance and biocompatibility, titanium alloy is widely used in the high-precision industries of human life, and is more applied in petroleum power generation, marine engineering and medical industries. As a not scarce resource on earth, titanium has not been widely used for many years, mainly due to the high application cost of titanium alloy.

With the rapid development and wide application of titanium alloys, the processing of titanium alloys is still a major industrial problem. Titanium alloy is a typical difficult-to-cut metal material. The main problems in the cutting process include: (1) small deformation coefficient, (2) high cutting temperature, (3) large cutting force per unit area, (4) the phenomenon of cold hardening is serious, (5) the tool is easy to wear. Therefore, it is of great significance to study thoroughly and systematically the turning temperature and turning vibration characteristics of titanium alloy during machining.

In recent years, many scholars at home and abroad have studied the temperature and vibration characteristics of titanium alloy turning from theories, experiments, finite element simulation, and predictive modeling.

2 Current status of cutting temperature

Due to its high strength and low thermal conductivity of titanium alloy, the contact area of the tool and the workpiece produces high pressure and temperature, while all the energy consumed is almost converted to thermal energy. Cutting heat is generally transmitted from chips, tools, work-pieces and surrounding media. The high temperature and strong vibration of the titanium alloy are cut due to its high strength, low thermal conductivity and low elastic modulus. Reasonable control of cutting temperature can reduce the impact of thermal deformation, reduce tool wear, extend the service life of tools, and thus stabilize the processing precision and processing quality.

2.1 Factors affecting the cutting temperature

Practice has proved that the workpiece material, tool material and structure, processing mode and cutting parameters setting are all important factors affecting the cutting temperature.

Davies [1] reviews several widely used cutting temperature measurements, Shows how the temperature can be monitored during material removal, in comparison with criteria critical for removal of measured materials, provide a reference to scholars in the field of work; Han [2] performed high speed cutting and milling tests on Ti-6Al-4V titanium alloy material, studied and analyzed the trend of cutting temperature under dry cutting, air jet and nitrogen jet; Zhou [3] used low-temperature cooling technology. The results show that low-temperature cooling can significantly improve the stress and thermal stress of cutting tools; Liu et al [4-7] combined with ultrasonic vibration vehicle cutting to study the changes in amplitude, cutting velocity and displacement in different directions, analyze the influence of the two

*Corresponding author e-mail: zsclsc@263.net

cutting methods on the cutting force and maximum temperature changes and material processing properties; Venkatesan [8] made an experimental study on laser assisted machining of Inconel 718 using turning process, and analyzed the effects of laser cutting parameters (i.e. cutting speed, feed rate and laser power) on cutting force and cutting temperature; Xavior [9] carried out precision turning of Ti-6Al-4V titanium alloy. Based on the experimental values, analysis of variance was carried out to understand the effects of cutting speed, feed rate, cutting depth and tool tip radius on cutting force, surface roughness and cutting tool temperature.

2.2 Use finite element simulation to monitor and predict cutting temperature

With the development of computer science and technology, finite element simulation becomes the main means of study heat and cutting temperature. He et al [10-11] used finite element analysis (FEA) technology to model titanium alloy, the influence of the cutting parameters on the cutting temperature is analyzed, a regression model of cutting parameters and cutting temperature was established to predict titanium alloy cutting temperature; Veeranaath [12] studied the changes of workability, cutting stress and temperature of PCBN tool in dry processing Ti-6Al-4V by FEA technology; Tian[13] performed numerical simulation of the high-speed milling temperature field of titanium alloy TC4, the influence of the cutting parameters on the front and rear blade temperature and its distribution is studied; Based on the fluent solid energy equation, Du [14] constructed the finite element model of heat transfer on the tool chip contact surface in the second deformation zone, and discussed the effects of coating material, coating thickness and tool chip actual contact area on heat transfer when the coated tool was cutting H13 hardened die steel; Yan [15] combined with FEA technology to establish a 3 D cutting finite element model of TC4 titanium alloy with three elements, we analyze and compare the cutting force of three different coating tools; He [16] uses FEA technology to establish a cutting model, the effect of the cutting parameters on the cutting temperature field and the residual stress field is analyzed for simulation; combined with the Johnson-Cook Law of Materials, Zhang [17] established a three-dimensional finite element model (FEM) for oblique cutting and studied the distribution of cutting force, temperature and residual stress during oblique cutting of TC21 alloy; Wang [18] uses two finite element software under the same simulation parameters to analyze the cutting temperature distribution and the trend of cutting temperature with the cutting speed.

2.3 Predictive modeling using mathematical methods

In addition to the simulation analysis of cutting temperature using finite element software, many scholars also use mathematical models to establish analytical models related to cutting temperature, and predict the change of cutting temperature during the cutting process, with more high efficiency.Qu [19] establishes the mathematical model of D406A CNC cutting temperature and cutting force and determines the parameters in the model by cutting test; Ulutan [20] and Yan [21] predict the temperature during processing based on a thermal modeling technology which can greatly reduce the calculation time without causing much impact on the accuracy. Mohring [22] proposes a numerical iterative method for calculating the temperature of the contact area between the chip and the tool.The temperature of the tool measured on the front and rear knife faces is compared with the calculated temperature.Shan [23] developed an improved analytical model based on the mobile heat source method to predict the distribution of the cutting temperature in the orthogonal cutting of the titanium alloy TC4, showing that the relative difference in the predicted temperature was 0.49% to 9%, indicating its high predictive power accuracy. Dayan [24] used the general full factor method to design the test, and established three prediction models of surface roughness, temperature and metal removal rate according to the obtained test data. The established model is in good agreement with the test results.Pervaiz [25] combines traditional finite element machining simulations with computational hydrodynamic models to establish a new method to analyze the temperature distribution of cutting tools to predict cutting heat and knife tip temperature during machining; Imbrogno [26] developed a new 3 D finite element model to predict cutting force, temperature and corresponding microstructural changes in semi-refined Ti-6Al-4V under dry low temperature, and calibrated and verified.

3 Current research status of cutting vibration

Titanium alloy causes sharp cutting vibrations during cutting due to its lower elastic modulus and high strength. When the cutting vibration becomes large to a certain extent, the tool vibration directly affects the tool wear situation, and by monitoring the tool vibration can obtain information about the tool wear, improve the production quality and efficiency of the cutting [27].To this end, scholars study the influence of cutting parameters on the cutting vibration, the structure of the machine tool itself, the structure of knives and other titanium alloy cutting methods.

3.1 Impact factors of cutting vibration

Practice has proved that the workpiece material, tool material and structure, processing mode and cutting parameters setting are all important factors affecting the cutting temperature.

Sun [28] performs the first systematic study of the numerical and theoretical aspects of the high-speed vibrational cutting process of Ti-6Al-4V, help to deepen the understanding of the vibration effect of metal cutting, provide practical guidance for retraining and utilization of vibration assistance processing; Yan [29] studied the change of tool vibration with cutting parameters during the processing of TA2 pure titanium and TC4 titanium alloy; Zhou [30] goes through external circular vehicle machining of TA15 titanium alloy, the vibration signal characteristic values were collected and extracted, the influence law of the cutting amount on the cutting vibration and the surface roughness is analyzed, explain the effect of the cutting amount on the vibration; Zhu [31] systematically investigated the relationship between tool wear, chip morphology, and cutting vibration, tool wear was found to aggravate the cutting vibration, thus aggravating the chip edge wear.

3.2 Monitor and predict tool wear based on cutting vibration

Reasonable monitoring of cutting vibration can reduce the zigzag cutting chips caused by titanium alloy with high strength and low heat conductivity, and aggravate the vibration of the tool system, and then weaken the wear of the tool and the deterioration of surface processing quality.

Scholars analyze the optimal processing conditions of titanium alloy by building a prediction model. Silva [32] proposes a model to predict milling force, which is suitable for milling of ring tool and considers the cutting vibration effect of processing. Combined with FEM simulation analysis software, Patil [33] performs two-dimensional FEM transient simulation and experimental characterization of titanium alloy under ultrasonic vibration assisted cutting condition using DEFORM software. Wang [34] uses the wavelet characteristics and support vector machine mode recognition algorithm to establish a milling vibration recognition system, and divide the vibration signal into three categories.

3.3 Measurements of reducing cutting vibration

The design of special structure knife rack, special shape tool and ultrasonic vibration auxiliary cutting methods can have a good effect on reducing cutting vibration. By manufacturing the hollow elements of special structure in the rack shaft, Vogel [35] greatly reduces the vibration amplitude of the tool and greatly improves the production quality of the piece. The texture tool proposed by Suzuki [36] can effectively cause process damping and inhibit vibration, which can play a good role in the experiments compared to ordinary tools. Rao [37] found that a new tool with a micro porous pattern on the front blade and side of the tool can reduce the friction of the front blade and thus reduce the vibration. Zheng [38] developed a rotating vibration control system based on multiple time delay, and numerical simulation found that the designed controller can effectively inhibit the cutting vibration and improve the stability of the cutting process. Identifying the dynamics of the cutting process based on working mode analysis, Kim [39] proposes a new vibration monitoring method that can determine the stability level of cutting before the vibration becomes unstable. Martin [40] studied the influence of cutting parameters on the vibration of machined parts. The experimental results show that the feed rate is the most important factor affecting the machining effect. Chen [41] proposes a novel nested artificial neural network that is more accurate to predict cutting states such as cutting vibration and surface roughness compared to conventional RSM and linear regression models. Compared to the traditional mathematical model in the new method, the newly proposed neural network class model has too high judgment coefficient and lacks the tests of the unfamiliar test set.

4 Conclusion

The temperature and vibration generated during the actual cutting process are coupled to each other and change over time. Based on the theoretical analysis, the scholars and scholars studied the temperature and vibration of titanium alloy cutting. However, the following points can be made to improve the cutting temperature and vibration.

(1) In terms of testing: At present, there are many research results on a single measurement method of cutting temperature and cutting vibration, and there are few research results on the simultaneous collection of cutting temperature and vibration. Therefore, in future research, a simultaneous measurement system for cutting temperature and vibration can be built, which provides an experimental basis for the analysis of the coupling characteristics between cutting temperature and vibration.

(2) Intelligent prediction: Most of the current research literature analyzes the influence of cutting parameters and tool structure parameters on cutting temperature and vibration through experimental testing and theoretical research, and uses a large number of numerical simulation methods such as finite element models to analyze temperature and vibration characteristics. Fewer scholars have established vibration and temperature prediction models, and more vibration and temperature prediction models can be established in the future to provide a reference solution for the current problem of the difficulty of measuring the turning temperature of titanium alloys in factories.

(3) In terms of coupling characteristics analysis: At present, most of the literature establishes the empirical formula of turning temperature on cutting parameters or the prediction model of cutting vibration on turning parameters. Few documents establish the fitting model between cutting vibration and temperature, and study the relationship between the two The coupling characteristics. Therefore, in the follow-up research, the research on the coupling characteristics between cutting temperature and turning vibration needs to be further strengthened.

Acknowledgments

This work was supported by the Postgraduate Research & Practice Innovation Program of Jiangsu Province (KYCX21_2618), Joint scientific research project of Sino foreign cooperative education platform in Jiangsu Province, the Science and Technology Plan Project of Xuzhou City (KC20188), and Undergraduate Innovation Training Program (202010320033Z).

References

1. J. Davim: Machining of Titanium Alloys (Springer Heidelberg New York Dordrecht London, England 2014)
2. M.L. Han, Y.M. Li and W. Zhao: Tool Engineering, Vol. 42 (2008) No.6, p.10.
3. B. Zhou, L. Hou, R. Tang and Q. Zhang: Combined machine tool and automatic machining technology, (2015) No.7, p.10.
4. A.J. Liu, Z.J. Zhou and B.L. Hu: Tool Engineering, (2017) No.12, p.40.
5. Z.T. Hu, N. Qin and F. Liu: Machinery Design & Manufacture, (2018) No.2, p.164.
6. Y.F. Gao, H. Chen and Y. Jiang: Journal of Nanchang University (Engineering Edition), Vol. 39 (2017) No.3, p.263.
7. Y.S. Xu, P. Zou and W. Wang: Journal of Northeastern University (Natural Science Edition), Vol. 38 (2017) No.1, p.95.
8. K. Venkatesan, R. Ramanujam and P. Kuppan: Procedia Engineering, Vol. 97 (2014) , p.1637.
9. M.A. Xavior and R. Vinayagamoorthy: Procedia Engineering, (2014) No.97, p.308.
10. Z.X. He, Y.H. Hu, C.J. Yin: Journal of Chinese Agricultural Mechanization, Vol. 35 (2014) No.4, p.124.
11. J.B. Sun, Y.H. Hu, L. He,: Machine Tool & Hydraulics, Vol. 44 (2016) No.5, p.68.
12. V. Veeranaath: Materials Today: Proceedings, (2018) No.5, p.19547.
13. X.M. Tian, J.G. Ren and J.H. Zhou: Tool Engineering, Vol. 46 (2012) No.10, p.37.
14. J. Du, T.T. Zhou and J.J. Zhang, et al: Tool Engineering, (2018) No.5, p.51.
15. S. Yan, M.J. Wang, X.K. Hu: Combined machine tool and automatic machining technology, (2014) No.10, p.135.
16. X.J. He, G.D. Chen and T. Nelias: Mechanical Science and Technology for Aerospace Engineering, Vol. 30 (2011) No.12, p.2116.
17. Y.C. Zhang, T. Mabrouki and D. Shen, et al: International Journal of Material Forming, Vol. 4: (2011) No.1, p.11.
18. H.F. Wang, T. Xiao and W.G. Wu: Machinery Design & Manufacture, (2012) No.09, p.48.
19. Z.X. Qu, and L.F.Zhang: NC [J]. Tool technology, Vol. 43: (2009) No.4, p.51-56.
20. D. Ulutan, I. Lazoglu and C. Dinc: Journal of Materials Processing Technology, Vol. 209 (2009) No.2, p.1111.
21. S.J.Yan, D.H.Zhu and K.J.Zhuang: Journal of Materials Processing Technology, Vol. 214: (2014) No.12, p.2985.
22. H.C. Möhring, C. Kushner and M. Storchak: CIRP Annals, Vol. 67: (2018) No.1, p.61.
23. C.W. Shan, X. Zhang and B. Shen, et al: Chinese Journal of Aeronautics, Vol. 32: (2019), p.759.
24. G.M. Dayan, R.H. Kishore, V. Praveen: Materials Today: Proceedings, Vol. 18: (2019), p.2191.
25. S. Pervaiz, I. Deiab and E.M. Wahba, et al: Procedia CIRP, Vol. (2014) No.17, p.750.
26. S Imbrogno, S Sartori, A Bordin, et al: Procedia CIRP, Vol. (2017) No.58, p. 475.
27. R. Kataoka and E. Shamoto: Precision Engineering, VoL. 55 (2019), p.322
28. S. Zongtao, S. Fei and M. Wei: International Journal of Mechanical Sciences, (2018) No.148, p.510
29. S.J. Yan, B.C. Xu and Q. Xin, et al: Manufacturing Technology & Machine Tool, (2018) No.8, p.129
30. P.P. Zhou, J.Q. Li and L.F. Xu, et al: Tool Engineering, VoL. 50 (2016) No.5, p.21
31. Z.J. Zhu, J. Sun and J.F. Li, et al: The International Journal of Advanced Manufacturing Technology, VoL. 83 (2016) No.9-12, p.1477
32. V.A.D. Silva, R. Pederiva and H.K. Neto, et al: The International Journal of Advanced Manufacturing Technology, (2018) No.94, p.2115
33. S. Patil, S. Joshi and A. Tewari, et al: Ultrasonics, VoL. 54 (2013) No.2, p.694
34. Z.G. Wang, S. Nakashima, M. Larson: Procedia CIRP, (2014) No.17, p.523.
35. F. A.M. Vogel, S. Berger, E. Ozkaya, et al: Procedia Manufacturing, (2019) No.40, p.32.
36. N. Suzuki, W. Takahashi, H. Igeta, et al: CIRP Annals, VoL. 69 (2020), p.93.
37. C.M. Rao, S.S. Rao and M.A. Herbert: Journal of Manufacturing Processes, VoL. 36 (2018), p.93.
38. P.X. Zheng and X.H. Long: Theoretical and Applied Mechanics Letters, VoL. 3 (2013) No.6, p.e063003.
39. S. Kim and K. Ahmadi: Mechanical Systems and Signal Processing, VoL.130 (2019), p.315.
40. S. Martín-Béjar, F. Trujillo, C. Bermudo, et al: Procedia Manufacturing, VoL.41 (2019), p.835.
41. Y. Chen, R.L. Sun, Y. Gao, et al: Measurement, VoL.98 (2017), p.25.

Development of a new multi-purpose wrench based on a lathe

Peng Wang [1], Yang Wang [1], Haotong Yu [1], Na Cheng [1], Honghua Hao [2] and Zhou Yan [3,*]

[1] School of Mechatronics Engineering, Shenyang Aerospace University, Shenyang, China
[2] School of Aeronautics and Astronautics, Shenyang Aerospace University, Shenyang, China
[3] Engineering Training Center, Shenyang Aerospace University, Shenyang, China

Abstract. In lathe operation, the most frequently used wrenches are chuck wrenches and tool holder wrenches, which are different in structure and function and cannot be replaced, and forgetting to remove the wrenches will cause safety accidents. To solve the above problems, this paper designs a new type of wrench, which integrates the functions of chuck wrench, tool holder wrench and various screwdrivers, and can easily change the form of wrench by certain operation, which improves the production efficiency and eliminates the safety hazards.

1 Introduction

In the machining workshop, the common wrenches for lathes are chuck wrenches and tool holder wrenches. If the operator forgets to remove them from the chuck or tool holder, the wrenches will easily fly out when starting the lathe, which will endanger the personal safety of the operator and cause damage to the turning tool and guide rail, thus causing safety accidents.

The chuck wrench and the tool holder wrench are each used in use and cannot be replaced, and it is necessary to change these two wrenches frequently, which is easy to take mixed; the two pairs of wrenches take up space and other resources when kept and stored.

In view of the above situation, it is necessary to design a kind of safety multi-wrench, which can combine the functions of chuck wrench, tool holder wrench and many kinds of screwdrivers in one, and the volume is only the size of one wrench, so that there will be no situation of taking the wrong wrench, and it can also reduce the equipment manufacturing cost and save space and resources.

On the other hand, in view of the most important shortcoming and deficiency of the existing chuck or tool holder wrenches, which cause a large number of accidents due to failure to remove them in time, a wrench that can automatically eject the chuck or tool holder is designed so that no accident occurs because the operator forgets to remove the wrench.

2 Main principle and structural design of safety chuck wrench

2.1 Main design principles and program descriptions

Forhow to realize the key problem ofautomatic ejection of the wrench from the chuck or the tool holder, our design solution is to make a structural design on the wrench body and install a kind of automatic ejection device, so that the wrench can automatically eject from the chuck or the tool holder when the operator no longer applies force to the wrench.

In this scheme, the spring force of the spring is used to realize the automatic ejection function of the wrench, and the reset function of the spring is also used to control and adjust the position and stroke of the work by the spring. Considering that the actual structure of this design is mainly cylindrical, and that cylindrical springs are the most widely used, standardized in manufacture and relatively simple in structure, it was decided to use cylindrical compression springs.

In order to realize the combination of two wrenches into one, the "outer sleeve inner core" type was adopted, and the chuck wrench was placed inside the sleeve of the tool holder wrench.

In addition, a screwdriver head slot is milled out at one end of the lever, so that different types of screwdriver heads can be replaced and installed; various types of replaceable screwdriver heads, such as one screwdriver head, Phillips screwdriver head, square screwdriver head, etc., are installed inside the lever to facilitate the replacement of screwdriver heads.

* Corresponding author: sdyanzhou16888@ edu.sau.cn

2.2 Overall structure and important structural design

2.2.1. Overall structure and working principle.

Figure 1 shows a schematic diagram of the structure of the chuck wrench body, taking the existing chuck wrench as a prototype, and the structure is designed as a step shaft.

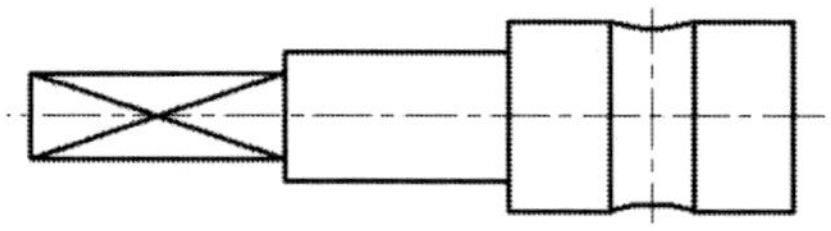

Figure 1. Schematic diagram of chuck trigger body structure.

The overall structure of the new multi-purpose wrench for lathe is shown in Figure 2, including the top of the wrench body, upper spring, slot, force lever, chuck wrench body, lower spring and tool holder wrench sleeve, using the "outer jacket inner core" type, combining the chuck wrench body and tool holder wrench sleeve together.
The chuck wrench body 5 and the tool holder wrench sleeve 7 are the main functional structures, the chuck wrench body 5 is installed inside the tool holder wrench sleeve 7, and the chuck wrench can extend out of the tool holder wrench sleeve 7 through the lower edge of the inner hole of the tool holder wrench sleeve 7 in the process of downward pressure, and the function of the chuck wrench and the tool holder wrench can be realized through the difference of the limit position; the tool holder wrench sleeve 7 plays the role of protecting the whole wrench body and the spring.

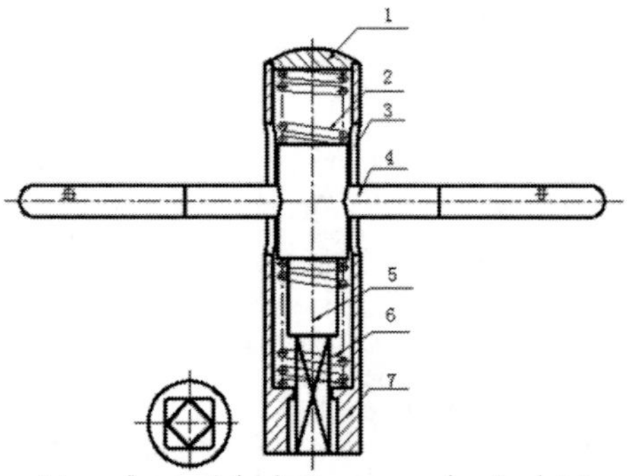

1.top of wrench body 2. upper spring 3. slot 4. application lever
5. chuck wrench body 6. lower spring 7. tool holder wrench sleeve

Figure 2. Overall assembly diagram.

The upper spring 2 and lower spring 6 are each installed at the upper and lower shoulders of the chuck wrench body 5, which can be telescoped during the downward and backward presses of the chuck wrench body 5, respectively, to realize the automatic reset of the wrench body.
The force lever 4 transversely passes through the slot 3 on the sleeve of the knife holder wrench and the chuck trigger body 5 and is fixed on the chuck trigger body 5, which can slide along the slot 3 and its downward and backward pressing position is restricted by the slot 3. The inner hollow of the force lever 4 can store the screwdriver bit, Phillips screwdriver bit, square screwdriver bit, etc.
The working principle is as follows: When the chuck tightens or loosens the working piece press hard on the force-applied lever 4, the lower spring 6 contracts under pressure and the chuck trigger body 5 extends, rotate the force-applied lever 4 hard to tighten or loosen it, after which the force pressed on the force-applied lever 4 is canceled and the force-applied spring 6 is reset, thus pushing the chuck trigger body 5 upward and thus making the chuck trigger body 5 eject from the chuck square hole.
When it is necessary to change the tool, let the bottom align with the screw on the tool holder, press hard on the top 1 of the trigger body, the upper spring 2 is contracted by the pressure, rotate the force-applying lever by force thus tightening or loosening the screw, after which the force pressed on the force-applying lever 4 is canceled and the pressured upper spring 2 is reset so that the trigger body is ejected from the screw.
When you need to use a screwdriver, remove the force lever and select the corresponding screwdriver head.

2.2.2. Structural design of the force bar.

By designing the inner hexagonal hole at one end of the force application rod, it makes it possible to change the required screwdriver head at any time; the other end is designed with a round hole to place different types of screwdriver heads. This structure way can collect a variety of screwdrivers into one, realizing the integration of a variety of screwdrivers and trigger body installation, which is convenient to use.

The structure of the force-applied rod is shown in Figure 3. Replaceable screwdriver heads of various types are installed inside the hollow force application lever, and the operator can take out the screwdriver head on its open side for use.

Both ends of the lever are knurled to provide aesthetic and anti-slip effects.

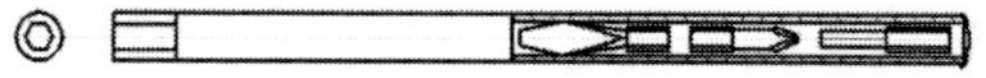

Figure 3. Schematic diagram of the structure of the force-applying rod.

2.2.3. Itinerary design.
In order for this improved design product to move within a certain stroke under the action of spring force and external force applied by the operator, and eventually achieve the protection function of automatic ejection, parts are needed to limit its stroke.
As can be seen in Figure 4 of the overall structure, a sleeve part is utilized. The sleeve is connected to the chuck body by two symmetrical 180°key slotted holes in the wall of

the sleeve, and the lever is passed through the holes, because the lever is fixed, so that the chuck body can only move within the range of its own key slotted holes. The length of the through-hole is the maximum travel of the wrench when it works, and its value is designed according to the overall structure and working condition of the wrench and the specific situation.
The stroke is not strictly required in this improved design, as long as it meets the working requirements of the whole designed product. The specific working stroke integrates the size of the chuck wrench head, the size of the tool holder sleeve and the actual situation at work, and the design stroke L=35mm.
When designing the length of the stroke hole, it is necessary to take into account that the force rod passes through the stroke hole, because the diameter of the designed force rod is 11mm, so the length of the stroke hole is designed as the actual maximum working stroke plus the diameter of the force rod, which is 46mm.

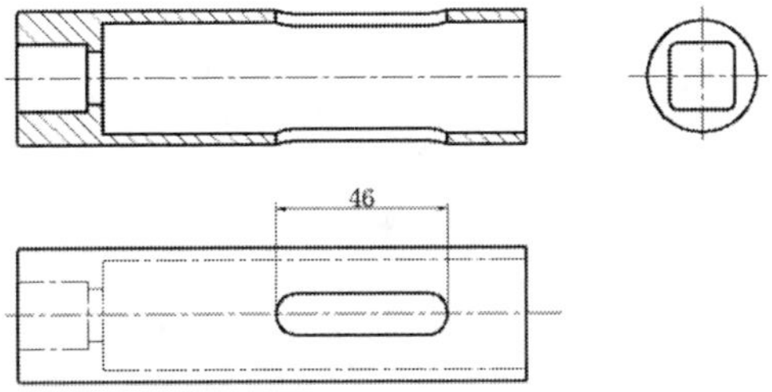

Figure 4. Tool holder wrench sleeve parts diagram.

2.2.4. Spring design.

The design of the spring is a key point of the new multi-use wrench. First of all, according to the work situation and specific conditions selected materials, the design, the use of the spring environment is room temperature environment, neither corrosive media nor bear a large impact load, so the use of general high-quality carbon steel can be selected after the material to determine the allowable stress. Spring material selection of 65 high-quality carbon steel, heat treatment process for oil quenching, 400°C~ 420°C tempering, hardness HRC42~ HRC48 can be. Because the spring steel wire diameter is less than 8~10 mm, so the use of cold roll method of manufacture, the spring ends and tight, grinding flat.

Figure 5. Physical drawing of the new dual-use wrench for lathe.

3 Conclusion

The new lathe multi-purpose safety wrench adopts reasonable structural design and improves the existing two types of wrenches, integrating the functions of chuck wrenches, tool holder wrenches and a variety of screwdrivers, and its volume is only the size of a wrench, with compact structure and convenient operation, so that the wrench will not be taken by mistake, saving the time of frequent wrench replacement, reducing labor intensity and improving production efficiency. More importantly, it avoids the safety hazards caused by forgetting to take off the wrench. It has a very significant effect on the long-term use of the equipment and the personal safety of the operator, etc. And the production cost is not high, which has a very great promotion value.

Acknowledgments
This work was financially supported by Shenyang Aerospace University Student's Platform for Innovation and Entrepreneurship Training Program fund. Project fund number: X202010143108.
First author: Peng Wang (2000.03--), male, Junior, the main research direction is aircraft manufacture engineering.
Corresponding author: Zhou Yan (1982.12--), male, engineer, main research direction is the application of numerical control technology and mechanical design research.

References

1. T.J. Luo, W.X. Pei, Design and application of safety chuck wrench, J. Machine Tools and Hydraulics, 2012,40(2):22-24
2. D.C. Gao,Application of safety devices in the improved design of lathe chuck wrenches, J. Heilongjiang science and technology information, 2015:44

https://doi.org/10.1051/matecconf/202135301018

3. X. Zhang, W.L. Pan,Design and application of safety protection device for lathe chuck wrench, J. Silicon Valley, 2013(7):181
4. H. Wang, J.G. Zhang, The development of lathe chuck locking safety wrench, J. Small and medium-sized enterprise management and technology, 2010:310
5. T.Zhang, L.Wu, Multifunctional lathe safety wrench, J. Science 24 Hours, 2007:35-36
6. J.H. Zhang, Improved design of chuck wrench, J. Equipment management and maintenance, 2011:35-36
7. Y.F. Zhou, Fundamentals of mechanical design. Beijing: Machinery Industry Press, 2008,pp. 84-332.
8. J.S. Wang, Mechanical general parts manual. Beijing: China Electric Power Publishing House, 2008,pp. 262-682.

Preparation and properties of recycled fine powder foam concrete

Yongmei Feng

Chengdu Polytechnic, Chengdu 610041 China

Abstr1act. With the rapid development of the construction industry brought by the accelerated process of urbanization, the problem of random stacking and landfill of waste concrete can be seen everywhere, which seriously affects the environmental health. The secondary utilization of waste concrete as recycled aggregate has become a research hotspot. This paper mainly studies the influence of the cementing material system of recycled fine powder foam concrete on the water absorption performance of recycled fine powder foam concrete, and explores the feasibility of using recycled fine powder to prepare foam concrete. The experiment also used recycled fine powder instead of some cement to prepare foam concrete. The effects of the amount of recycled fine powder, the amount of hydrogen peroxide and the ratio of water to binder on the compressive strength, water absorption, thermal conductivity and microstructure of the foam concrete were studied. The results showed that the water absorption rate of recycled fine powder foam concrete increased with the increase of red brick powder content, but the water absorption rate gradually increased. When the content of hydrogen peroxide is 6%, the ratio of water to binder has a greater influence on the thermal conductivity of foam concrete. When the amount of hydrogen peroxide is higher, the influence of water binder ratio on thermal conductivity decreases.

key word: Regeneration powder; foam concrete; thermal conductivity; hydrog en peroxide solution.

1 Introduction

Blister concrete, also known as foamed concrete, belongs to a kind of porous concrete. It is made of foam made from aqueous solution of foam agent and added into cementitious materials, sand, water and admixture, and after being evenly stirred, the lightweight porous building materials formed by pouring and curing are mainly made of cement sand foam concrete, lime cement sand foam concrete and so on [1]. The prominent feature of blister concrete is that there are a large number of foam holes inside, which not only reduces the quality of foam concrete, but also keeps heat insulation. At present, the main uses of foam concrete are to use the energy saving wall materials in the north, and to make use of foamed concrete to make light wall panels, blocks and cast-in-situ roofing insulation layer [2]. However, in recent years, we have been upgrading and optimizing the properties of foam concrete, and its application scope has gradually expanded. Foam concrete is used to fill dense large cavities. Generally, holes can be used for sewer wells or subway stations that are no longer used, or for illegal construction of cellars and basements [3]. This is mainly based on the low cost and high flow self compacting characteristics of foamed concrete. By filling the foamed concrete into the reverse side of the pier to reduce the lateral pressure of the pier on the bridge structure, it can also help to reduce the settlement. Backfill foam concrete can be used to balance the foundation settlement in road repair. Foam concrete has good thermal insulation and fire resistance, and the foam concrete produced by special cement, such as high alumina cement instead of ordinary cement, has the highest resistance to 1450 degree high temperature testing, so it can be applied to high temperature furnaces. Because foam concrete has a good energy dissipation function, it can be made into the target wall in the shooting range, and it can also be applied to the underground blast resistant tunnel structure .

Construction waste recycled powder is in the use of construction waste preparation of recycled aggregate, crushing classification, screening and finishing process, will bring about the content of about 15% - 20% of the total raw material, the average particle size is not more than 0.15mm of fine particles is called recycled powder. Due to the small particle size and light weight, it is easy to be affected by the air, temperature and other environmental factors and fly everywhere, which makes the environment seriously polluted. However, through chemical analysis and XRD analysis, the main chemical composition of the surface is cement slurry powder, and the main mineral phase is silica, as well as active components alumina and calcium oxide, so it is a kind of material with potential activity. At present, most of the

* Corresponding author: fengym19720315@163.com

utilization of waste concrete in China is concentrated on recycled aggregate, but the research and utilization of recycled fine powder produced in the process of crushing and shaping is still very few, so it is of great significance to study how to make good use of recycled fine powder.

Many experts and scholars have made great achievements in the research of recycled powder. Some studies have shown that the activity index of recycled micro powder is about 60% - 70%, which is similar to that of fly ash [4]. Boudali [5] et al. found that the heat treatment temperature was proportional to the activity of the regenerated powder, and the activity reached the peak at 600 ℃. Markin 's research shows that in addition to physical means to stimulate the activity of recycled powder, chemical activator can also be added to the recycled powder to stimulate the activity, so as to effectively improve its activity index [6]. However, there are differences in how to stimulate each chemical activator, the effect after stimulation and the optimal dosage. Many scholars have studied the compressive strength of recycled micro powder concrete, in which the content of micro powder is generally concentrated in 5% ~ 30%, and the maximum is 40% [7]. When the content of micro powder is less than 10%, the compressive strength of concrete is slightly higher than that of concrete without micro powder. When the content is less than 20%, the strength does not decrease obviously; However, when the content exceeds 30%, the strength decreases by about 20.2%. Jiang [8] used plate test to test the early crack resistance of recycled micro powder concrete products. The test conclusion is that when recycled micro powder is added to the concrete sample, the initial cracking time is effectively delayed, and the maximum crack width and length are also significantly shortened, which proves that recycled micro powder can significantly reduce the probability of early cracking of concrete.

The research team will use the red brick powder in the recycled micro powder as the main admixture to replace some cement to prepare the foam concrete, and then study its performance. The maximum use of red brick powder can also meet the relevant specifications of the foam concrete. Mainly from the following aspects. First, different proportion of recycled fine powder is used instead of cement to prepare foam concrete. Then the 28d compressive strength, mass water absorption and dry apparent density are measured, and the best amount of recycled fine powder can be determined at the same time when the recycled micro powder is used as much as possible. The regenerated micro powder foamed concrete prepared by different amount of foam stabilizer and the amount of foaming agent was studied. The changing trend of the related properties was tested and compared, and the best mixing amount was found respectively to determine the best mixture ratio. The influence of the amount of recycled fine powder on the thermal conductivity and water absorption of foam concrete is studied.

In view of the limited application of recycled fine aggregate, based on the activity and filling property of recycled fine aggregate and micro powder, the low thermal conductive foamed concrete was prepared by regenerating micropowder instead of some cement. The effects of regenerated fine powder, water binder ratio and H202 content on the mechanical properties, thermal conductivity and water absorption of the foam concrete were studied. The results showed that as the proportion of recycled fine powder increased, the water absorption of foam concrete increased, dry apparent density, 28d compressive strength and thermal conductivity decreased significantly.The specific research route is shown in Figure 1.

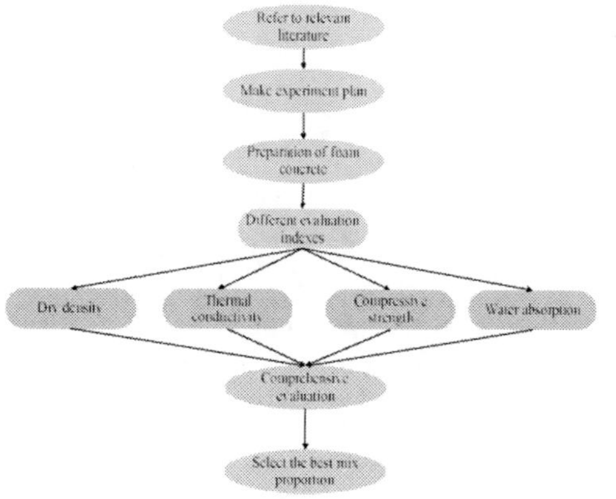

Figure 1 Technology roadmap of recycled fine powder foam concrete

2 Preparation of recycled fine powder foam concrete

2.1. Production of test object

In this study, we use mortar mixer to prepare recycled fine powder foam concrete. First, we use wet cloth to wet the mixing pot and stirring leaves, and then weigh the cement, red brick powder, foam stabilizer and fiber accurately through accurate electronic scales. Manually stir for 1-2 minutes, then slowly stir for 1-2 minutes with the machine, then add the water in the constant temperature water bath for 1-2 minutes at low speed, finally add the foaming agent, stir for 10 seconds at high speed, quickly pour into the test mold, maintain the test object according to the relevant specifications, demould it after 24 hours, and then put it into the standard curing room for performance test after curing for corresponding time. Figure 2 is a flow chart for the preparation of recycled fine powder foam concrete.

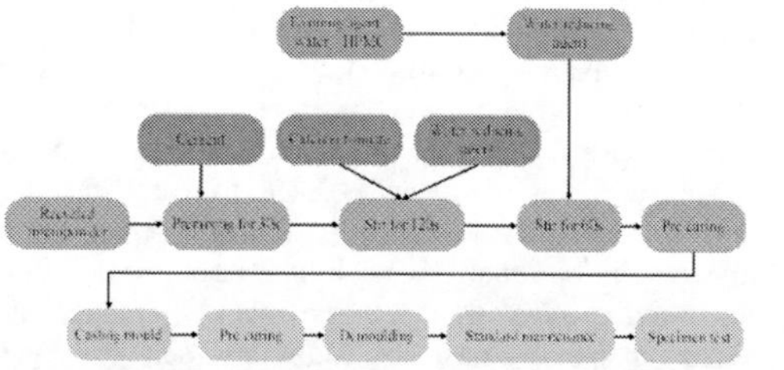

Figure 2 Flow chart of specimen production

 https://doi.org/10.1051/matecconf/202135301019

2.2 Mix proportion of recycled fine powder foam concrete

The ratio of water to material has a great influence on the preparation and performance of recycled micro powder foamed concrete. When the water is relatively small, the cement paste is over dried, and the amount of water required for red brick powder is greater than that of cement, making it easy to form agglomerate when the slurry is stirred, which leads to the fact that the mixture cannot be evenly mixed. Secondly, when the ratio of water to material is too small, the slurry is too thick. When the foaming agent is added for quick stirring, it cannot be evenly dispersed to all parts of the slurry, and even the foaming agent cannot enter into the slurry, resulting in the failure of the test. When the ratio of water to material is small, the resistance of gas produced by foaming agent to slurry is too large, which makes foaming difficult. The height is too low to reach the predetermined density level, and the distribution and size of pores are different. When the ratio of water to material is too large and the slurry is too thin, the gas produced by foaming agent will escape due to the small resistance, resulting in foaming phenomenon. At the same time, with the increase of the ratio of water to material, the decrease of the strength and dry density of recycled micro powder foam concrete will also accelerate. This is mainly because the main source of strength of recycled micro powder foamed concrete is cement, and the volume of cement will decrease as the ratio of water to material increases, and the strength will not reach the expected goal.

3 Study on properties of foam concrete

3.1 Influence of red brick powder content on properties of foam concrete

Dry apparent density is one of the most important indexes to evaluate the performance of foamed concrete. Therefore, fly ash is almost the first choice when selecting cementitious materials to replace foam concrete. The main reason is that there is a certain amount of glass in the fly ash after grinding, and the glass is composed of a large number of amorphous silica and aluminum oxide, which has the ability of depolymerization, which can not only improve the working performance of concrete products, but also improve their compressive strength. Therefore, the performance of fly ash is directly proportional to its internal vitreous content. The higher the vitreous content of fly ash, the better its performance. However, fly ash cannot be used as cementitious material alone, because fly ash itself cannot have the same cementitious effect as cement, without the addition of other admixtures. After adding fly ash into cement, it begins to hydration and forms a large amount of Ca $(OH)_2$. The active substances such as Al_2O_3 and SiO_2 in fly ash can continue to react with it to form secondary hydration. The final hydration products, calcium aluminate hydrate and calcium silicate hydrate, can improve the strength of concrete products. Recycled fine powder is mainly obtained from waste concrete or waste brick and tile after crushing and grinding, which contains a large amount of unhydrated cement slurry powder, a small amount of fine particles of cement stone and limestone, and some C2S, C3S and SiO2, so there is a certain possibility of activity. The main material used in this study is recycled red brick powder, whose main chemical composition is basically the same as fly ash and cement, but its raw material is mainly clay, so the content of SiO2 is relatively high. The related research also shows that the activity index of the milled regenerated powder is more than 65%.

Now the cementitious material system of recycled micro powder foamed concrete has basically been mixed with fly ash and regenerated fine powder. In this experiment, single red brick powder is used instead of cement to prepare recycled foam concrete. At the same time, single fly ash is selected as the control group, which can maximize the utilization of construction waste to regenerate fine powder. To achieve the purpose of reducing costs and resource utilization. In order to make full use of recycled micro powder as much as possible, it is also necessary to ensure that the performance of recycled fine powder foam concrete meets the relevant specifications. In this experiment, the cement content was fixed 60% to 100%, and the red brick powder was replaced by 0% to 40%. The performance of the foam concrete under different red brick powder content was tested, and the optimal dosage was selected by comparing the performance. The amount of foaming agent is 4%, the amount of foam stabilizer is 1.2%, the amount of fiber is 0.1%, the water material ratio is 0.6, and the test water temperature is 30 ± 1 °C. The amount of other materials is the ratio of cementitious materials. The test mix proportion is shown in table 1.

Table 1. Test mix proportion of recycled powder

Experiment number	Cement%	micro powder%	Water material ratio	foaming agent%
P1	100	0	0.5	7
P2	90	10	0.5	7
P3	80	20	0.5	7
P4	70	30	0.5	7
P5	60	40	0.5	7
P6	50	50	0.5	7
P7	40	60	0.5	7

From Figure 3, we can see that the water absorption rate of recycled micro powder foam concrete increases with the increase of red brick powder content, but its water absorption rate increases gradually. When the red brick powder is not added, the water absorption of pure cement foam concrete is 14.3%, which is the lowest among the 7 groups. When the content of red brick powder reaches 60%, the water absorption increases by 4%. It can be concluded that the increase of red brick powder has little effect on the water absorption of recycled fine powder foam concrete. From Figure 3, we can also know that the dry bulk density of the foam concrete in the pure cement group is the largest, and the 28d compressive strength is also significantly higher than that of other groups. From this we can know that mixing the red brick powder in the

 https://doi.org/10.1051/matecconf/202135301019

foam concrete will reduce the 28d compressive strength and dry apparent density, and the main source of strength of the foam concrete is the hydration product produced by the hydration reaction of cement. Therefore, with the increase of red brick powder, the amount of cement will be reduced, the water cement ratio is different, the volume of cement per unit volume will be reduced, and the activity of red brick powder is lower than that of cement. Therefore, the proportion of red brick powder replacing cement will increase, and the compressive strength of recycled concrete foam concrete will decrease accordingly.

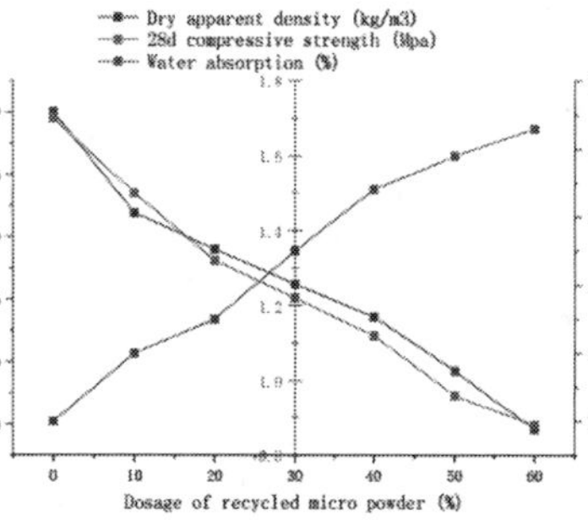

Figure 3 Effect of micro powder on the properties of foam concrete

The recycled micro powder foam concrete contains a large number of closed and disconnected stomata, which has excellent waterproof performance. However, the cementitious materials used in the preparation of foam concrete, such as cement, fly ash, red brick powder and so on, are fine particles. Foam concrete is mainly used as insulation material in the north. The water absorption rate leads to the decrease of freeze-thaw resistance, which limits the application and popularization of recycled fine powder foam concrete. In this experiment, the water absorption rate of recycled fine powder foam concrete increases with the increase of red brick powder content, and also because the red brick itself after crushing and grinding, the quality of the fine powder becomes lighter, the internal gap is large, and the water absorption of the material itself is large. However, the difference of water absorption between the seven groups is less than 5%, and the change of water absorption is related to the compressive strength, dry apparent density and the amount of admixture.

3.2 Research on thermal conductivity of foam concrete

Thermal conductivity is one of the most important thermal and moisture properties of building materials, which is closely related to building energy consumption, indoor environment and many other thermal and moisture processes. The thermal conductivity refers to the heat transferred through an area of 1 square meter in 1 second when the temperature difference between the two sides of a 1 m thick material is 1 degree under the condition of stable heat transfer. Foamed concrete has good thermal performance due to its large number of enclosed small spaces. It can be widely applied to interior and exterior insulation materials of houses. In this section, the thermal conductivity of foam concrete is explored in terms of the amount of recycled fine powder, water binder ratio and hydrogen peroxide.

Figure 4 shows the influence of the replacement rate of recycled micro powder on the thermal conductivity of foam concrete with different proportions of H2O2. By Figure 4A, the thermal conductivity of foam concrete decreases with the increase of the amount of recycled fine powder. The reason is that the coarse sugar on the surface of recycled fine powder has a large porosity. When the amount of recycled fine powder increases, the thermal conductivity of foam concrete decreases. From the Figure 4B, it is found that the thermal conductivity of foam concrete increases first and then decreases with the increase of the amount of recycled fine powder. The reason for this phenomenon is that the coarse particle size of recycled fine powder changes the pore structure of foam concrete and leads to the decrease of thermal conductivity. From Figure 4C, it can be found that when the water gel is relatively low, the replacement rate of recycled fine powder increases, and the thermal conductivity of foam concrete decreases. When the water gel is relatively high, the amount of recycled fine powder has little effect on the thermal conductivity of foam concrete. Compared with the three diagrams, it is found that the influence of the amount of recycled fine powder on the thermal conductivity of foam concrete decreases when the hydrogen peroxide content is increased from 3% to 9%.

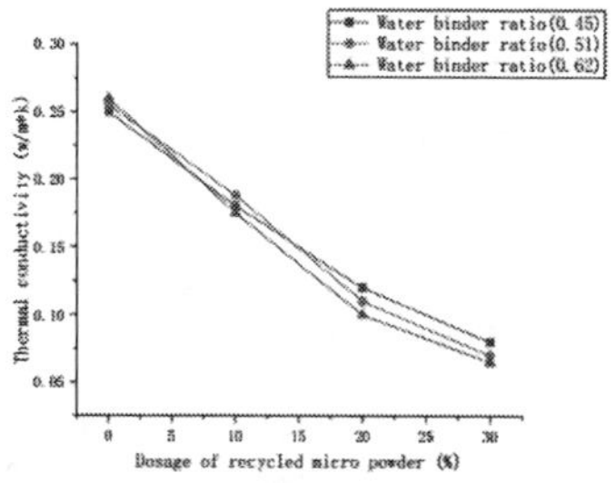

A. H_2O_2 dosage 3%

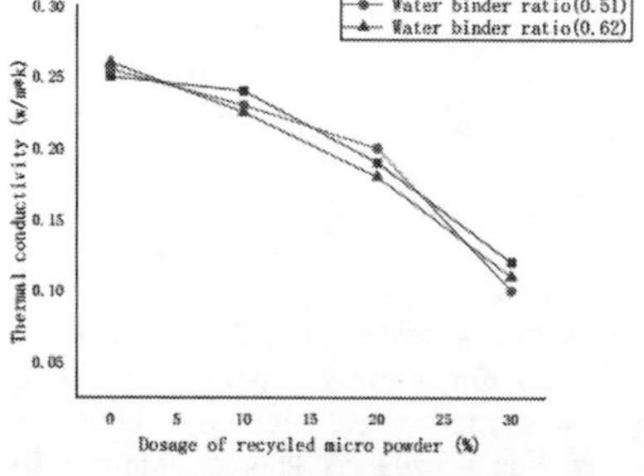

B. H_2O_2 dosage 6%

https://doi.org/10.1051/matecconf/202135301019

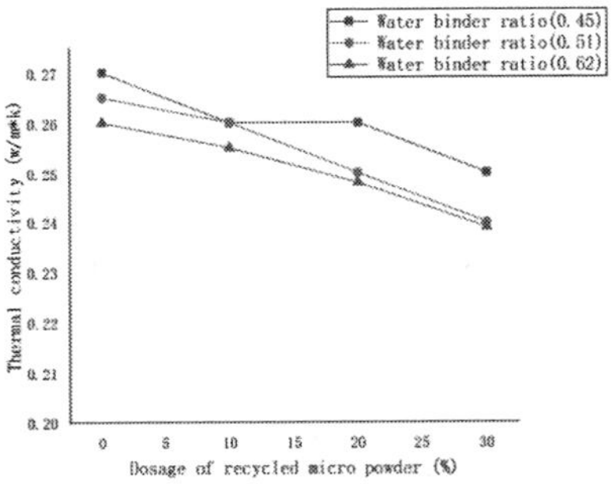

C. H_20_2 dosage 9%

Figure 4 Effect of recycled powder content on thermal conductivity of foamed concrete

4 Conclusion

In this paper, the preparation technology, performance and application of recycled micro powder foamed concrete are studied. The feasibility of preparing foam concrete with red brick powder in recycled fine powder is studied, and the best cementing material composition is found by changing the amount of red brick powder instead of cement. The proportion of red brick powder instead of cement is increasing. The dry density of recycled fine powder foam concrete will gradually decrease, and the compressive strength will also decrease, and its water absorption rate will increase slightly. Under the principle of making use of foam concrete as much as possible, cementitious material system (cement: red brick powder =60:40) is selected as the best amount of red brick powder. The prepared dry powder foam has a dry apparent density of 502kg/m3, a compressive strength of 1.43Mpa, and a water absorption of 17.1%.

When the content of hydrogen peroxide is 3% and 6%, the replacement rate of recycled fine powder increases with the addition of regenerated micropowder, and the thermal conductivity of foam concrete is lower. When the amount of hydrogen peroxide is 9%, the replacement rate of recycled fine powder has little effect on the thermal conductivity of foam concrete. When the water binder ratio is 0.62, the thermal conductivity of foamed concrete with recycled micro fly ash is the lowest. When the content of hydrogen peroxide is 6%, the ratio of water to binder has a greater influence on the thermal conductivity of foam concrete. When the amount of hydrogen peroxide is higher, the influence of water binder ratio on thermal conductivity decreases.

References

1. Canazza, M A, et al. "New Applications of Tire Powder Recycled by Micro-Waves."Materials Science Forum, vol. 1012, pp. 84-88, 2020.
2. Taherkhani, H, et al. "Investigating the properties of asphalt concrete containing recycled brick powder as filler. " European Journal of Environmental and Civil Engineering, vol. 1, pp. 1-11, 2020.
3. Kasuga, H, et al. "Influence of Addition of Emulsion Polymerized PTFE Powder on Particle Morphology and Mechanical Strength of Recycled PTFE Compacts. " Journal of the Society of Materials Science, vol. 68, pp. 62-67, 2019.
4. Yu, Jiangtao, et al. "Deformability enhancement of fiber-reinforced cementitious composite by incorporating recycled powder. " Journal of Reinforced Plastics and Composites, vol. 39, pp. 119-131, 2019.
5. Boudali, S, et al. "Influence of Fine Recycled Concrete Powder on the Compressive Strength of Self-Compacting Concrete (SCC) Using Artificial Neural Network." Sustainability, vol. 13, pp. 3111, 2021.
6. Markin, Nerella, et al. "Material Design and Performance Evaluation of Foam Concrete for Digital Fabrication. " Materials, vol. 12, pp. 2433, 2019.
7. Wei, SW, et al. "Mechanical properties of lightweight foam concrete filler for roadbed of high-speed railway. " Arabian Journal of Geosciences, vol. 14, pp. 1-10, 2021.
8. Jiang, C, et al. "Recycling of waste ceramic foams as fine aggregates in pervious concrete. " RSC Advances, vol. 10, pp. 2364-2367, 2020.

MATEC Web of Conferences 353, 01020 (2021)
AMME 2021
https://doi.org/10.1051/matecconf/202135301020

Research on the influence of heat treatment temperature on bearing capacity of high steel grade pipeline during in-service welding

Yan Xu [1], Yinglai Liu [1], Xianghui Nie [1], Zhenjun Feng [1], Liang Li [1], Fengping Yang [1], Jasmine [2]

[1] Tubular Goods Research Institute of CNPC, Xi'an, China
[2] Companhia Brasileira de Metalurgia e Minerao, Brazil

Abstract. During the welding of in-service pipeline, natural gas is continuously transported in the pipeline, which maintains a high gas pressure. Therefore, the welding process is completed under strong cooling conditions, and welding delay crack is easy to occur. Preheating before welding and heat treatment after welding can effectively control the hardened microstructure, reduce the residual stress and ensure the welding quality. In this study, the influence of heating temperature on the bearing capacity of high steel grade pipe during in-service welding repair was studied. The high temperature tensile test was used to simulate the bearing capacity of the pipe under heating and high temperature environment. It is found that when the heating temperature is below 400℃, the pipe strength remains at the original level. With the increase of heating temperature, when the test temperature is higher than 400℃, the yield strength and tensile strength of the pipe decrease significantly. When the test temperature is 450℃, the yield strength and tensile strength of the material decrease by 15.8% and 11.1%, respectively, compared with the normal temperature, which are lower than the pipe standard requirements. Therefore, it is suggested that when the heat treatment temperature is higher than 400℃ during in-service welding repair, it is necessary to consider reducing the pipeline pressure. At the same time, the box furnace heat treatment method was adopted to heat treat the pipe, and the tensile properties of the pipe after heat treatment were tested to analyze the bearing capacity of the pipe after heat treatment. It is found that when the heating temperature is higher than 700℃, the tensile properties of high grade pipeline steel pipe decrease sharply. It is suggested that when the heat treatment temperature is higher than 700℃, the risk assessment of the service safety of the heat treated pipeline should be carried out.

1 Introduction

In-service welding repair is one of the main repair methods in pipeline defects and leakage repair work. It can not only ensure the continuous pipeline transportation, but also reduce the degree of environmental damage, shorten the repair cycle and reduce the repair cost [1-2] . It has broad development prospects. At present, the in-service welding repair technology is faced with two technical difficulties, namely, burn-through and hydrogen-induced cracks [3-6] Among them, burn-through mainly considers that the welding arc is instantaneously loaded on the pipe metal in the welding process, and the increases instantaneously. When the inner wall temperature of the pipe reaches 982 °C, the local strength loss of the pipe metal will occur. Hydrogen-induced cracks mainly consider the continuous transmission of natural gas in the pipeline during the welding of the in-service pipeline, so the welding process is completed under strong cooling conditions, which is easy to produce welding delayed cracks. Preheating before welding and heat treatment after welding can effectively control the hardened structure, reduce the residual stress and ensure the welding quality. However, at present, the mainstream technology of pipeline steel production is to prepare high strength and toughness steel through low-carbon micro-alloying combined with thermo-mechanical control process. In practical application, when the heat treatment temperature is not controlled, the mechanical tensile properties of the pipe will change [7-10]. Therefore, it is necessary to study the influence of heat treatment temperature on the bearing capacity of high-grade pipeline during welding of in-service pipelines.

* Corresponding author: xuyan008@cnpc.com.cn

2 Mechanical tensile properties of high grade steel pipes at high temperature

2.1 Test Materials and Methods

The test material was taken from φ1016mm×12.8mmX80 steel tube, and the chemical composition is shown in Table 1. The transverse rod-shaped specimen was taken at the pipe body for high temperature tensile test. The diameter of the gauge section of the specimen was 6.25mm, and the length of the gauge section was 25mm. The test equipment is CMT5105 material testing machine, and the test standard is ASTM E21-09.

Table 1. Chemical composition of test steel tubes（Wt× 100%）

composition	C	Si	Mn	P	S	Cr	Mo	Ni	Nb	V	Ti	Cu
content	0.049	0.19	1.8	0.0072	0.0023	0.31	0.24	0.0084	0.088	0.0063	0.012	0.012

2.2 Test result

High temperature tensile test results are shown in Figure 1. When the test temperature is below 400°C, the strength index of the material fluctuates slightly. With the increase of test temperature, when the test temperature is higher than 400°C, the yield strength and tensile strength of the material decrease significantly. When the test temperature is450°C, the yield strength of the material is 533MPa, the tensile strength is 618MPa, compared with the room temperature tensile properties decreased by15.8% and 11.1%, the value is lower than the standard requirements. Therefore, it is suggested that when the post-weld heat treatment temperature is higher than 400°C, the depressurization treatment should be considered or the risk of in-service welding should be evaluated.

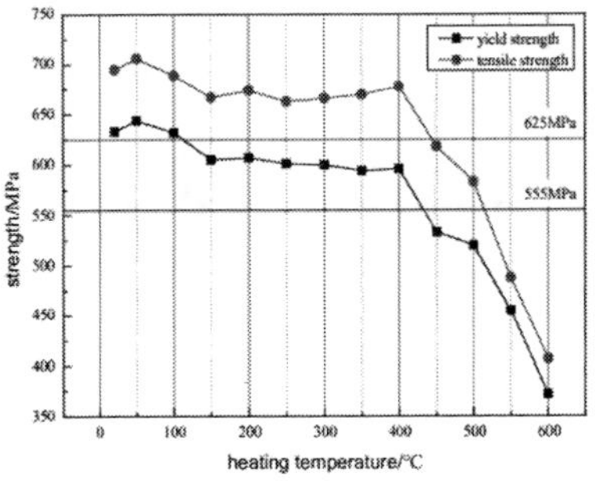

Figure 1. High temperature tensile test results

3 Mechanical tensile properties of high grade steel pipe after heat treatment

3.1 Test Materials and Methods

Four kinds of steel grade pipeline steel plates with different wall thicknesses were selected for heat treatment test. The chemical composition of the test material is shown in table 2. The heat treatment process is as follows: the test plate is heated to 300°C, 500°C, 600°C, 700°C, 900°C, 950°C, 1000°C and 1050°C respectively by box resistance furnace, and the furnace is cooled to room temperature after holding 55 minutes.

Table 2. Chemical composition of test steel tubes（Wt× 100%）

no.	steel grade	Thickness mm	C	Si	Mn	P	S	Cr	Ni	Mo	Nb	V
1#	X80	43	0.049	0.27	1.57	0.007	0.0013	/	0.21	0.23	0.045	0.025
2#	X70	40	0.066	0.22	1.5	0.013	0.0012	/	0.22	0.2	0.037	0.02
3#	X60	36	0.063	0.16	1.29	0.011	0.0025	0.18	0.2	0.19	0.041	/
4#	X60	45	0.1	0.27	1.22	0.01	0.0035	0.39	0.01	0.15	0.024	0.04

The rod-like tensile samples were taken at the center of the wall thickness of the raw material and the heat-treated test block, respectively. The diameter of the gauge section was 12.7mm and the length of the gauge section was 50mm. The tensile test was carried out on the UH-F500 KNI tensile testing machine according to ASTM A370-2012a standard.

3.2 Test result

The yield strength of pipeline steel after heat treatment is shown in Fig. 2. After heat treatment at 300°C – 700°C, the yield strength of pipeline steel increases slightly, which is mainly related to the precipitation strengthening of steel after heat treatment. In the temperature range of 700°C – 900°C, the material enters the temperature range of ferrite and austenite, and the original TMCP controlled rolling structure changes to form partial recrystallized structure. The yield strength decreases with the increase of heat treatment temperature. When the heat treatment temperature was higher than 900°C, the buckling strength of the material increased with the increase of heat treatment temperature. The grain grew up and the yield strength began to increase after all the austenitized microstructure was cooled by air, making the two-phase region at 900 °C the lowest point of yield strength. The yield strength of sample 1 decreased by 94MPa, that of

https://doi.org/10.1051/matecconf/202135301020

sample 2 decreased by 161MPa, that of sample 3 decreased by 212MPa, and that of sample 4 decreased by 191MPa. The higher the raw material strength was, the greater the yield strength decreased. When the heat treatment temperature rises to 1050°C, the yield strength of No.2, No.3 and No.4 materials increases rapidly due to the further growth of grains.

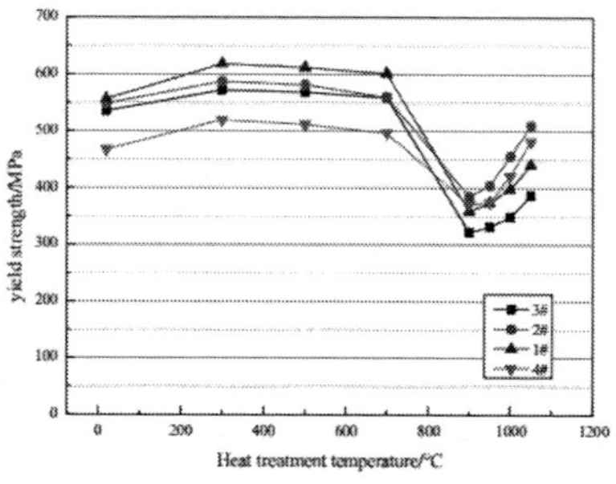

Figure 2. Yield strength of pipeline steel after heat treatment

The tensile strength of the pipeline steel after heat treatment is shown in figure 2. When the heating temperature is not higher than 500°C, the change of the tensile strength of the material is not obvious. With the further improvement of the heating temperature, the tensile strength of the pipeline steel decreases first and then increases, and reaches the lowest value in the temperature range of 900°C ~ 950°C. The tensile strength of sample 1 is reduced by 194MPa compared with the measured value of raw materials, sample 2 is reduced by 148MPa, sample 3 is reduced by 155MPa, and sample 4 is reduced by 191MPa.

By comparing the mechanical tensile properties, it is suggested that when the heat treatment temperature is higher than 700°C, the service safety of the pipeline after heat treatment should be evaluated.

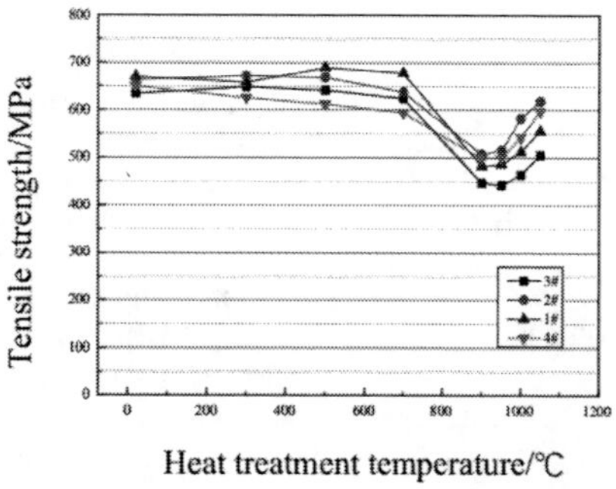

Fig 3 Tensile strength of pipeline steel after heat treatment

4 Conclusion

(1) The high temperature tensile test was used to simulate the bearing capacity of the pipeline under high temperature heating environment. It was found that the strength of the pipe maintained at the original level when the heating temperature was below 400°C. With the increase of heating temperature, when the test temperature was higher than 400°C, the yield strength and tensile strength of the pipe decreased significantly. When the test temperature was 450°C, the yield strength and tensile strength of the material decreased by 15.8% and 11.1% compared with those at room temperature, respectively, which were lower than the requirements of the pipe standard. It is suggested that the pipeline pressure should be reduced when the heat treatment temperature is higher than 400°C.

(2) The box furnace heat treatment method was used to heat the pipe, and the tensile properties of the pipe after heat treatment were tested. It was found that when the heating temperature was higher than 700°C, the tensile properties of high grade pipeline steel pipe decreased sharply. It is suggested that when the heat treatment temperature is higher than 700°C, the service safety of the pipeline after heat treatment should be evaluated.

References

1. WU Linjun. Reliability Analysis of Repair Technology of Pressure Pipeline in Service [J]. Quality and Technical Supervision Research, 2020(06):33-36.
2. HAO Jianbin, WU Xinjuan et al. In-Service Pipeline Repair and Simulation Testing Method of Emergency repair and Welding Process of Pipeline. OGST, 2006(02):37-40+62+2.
3. HAN Dehui. Burn Through Wall Thickness Phenomenon Analysis During Welding on Service Pipeline [J]. Welded Pipe and Tube, 2006(01):30-31+78.
4. LU Yu-xiu, LI Qiang, CAO Yong-feng. Study on Changing Law of In-Service Welding Pool Size of Oil and Gas Pipelines [J]. Pipeline Technique and Equipment,2021(03):59-62.
5. CHEN Juan, JI Feng, LIU Xuebin, et al. Control measures of in-service pipeline welding hydrogen-induced crack under low temperature environment [J]. Oil & Gas Storage and Transportation, 2014,33(12):1297-1300.
6. FU Wei, LYU Yuling. Burn-though and Hydrogen Induced Cracking Prediction of Oil and Gas Pipeline during In-service Welding [J]. Hot Working Technology,2018,47(23):241-245.
7. XU Yan, LIU Ying-lai, FENG Zhen-jun, et al.. Effect of Heat Treatment Temperature on Mechanical Tensile Properties of X80 Pipeline Steel [J]. Heat Treatment Technology and Equipment ,2020,41(06):37-41.

8. Liu Yinglai, Wang Gaofeng, Nie Xianghui, et al..Influence of Normalizing Temperature on Microstructure and Mechanical Properties of Controlled Rolling Steel for Preparation of Manifolds [J]. Natural Gas and Oil, 2015,33(05):79-83+12.
9. XU Minxu, LUO Lihui, ZHOU Huiping, et al. Influence of B -type Sleeve Welding on the Performance of X80 Pipeline Steel [J]. Materials Protection, 2019,52(06):149-156.
10. XU Yan, LIU Ying-lai, NIE Xiang-hui, et al. Influence of Process Parameters on the Forming Quality of L360M Steel Grade with Small Bending Radius Bend [J]. Heat Treatment Technology and Equipment, 2020,41(04):16-20.

Analysis and optimization of positioning scheme for machining 2.5mm groove

Huan Yu [1,2,*]

[1] School of Shenyang Institute of technology, Liaoning, China
[2] Liaoning Key Laboratory of physical information fusion and intelligent manufacturing of CNC machine tools, Liaoning, China

Abstract. The front bracket of a certain type studied in this subject is a part of the braking system and plays an important role in the braking system. The main parts of the front support are grooved surface, 44.3mm cylindrical surface, M16 threaded hole plane, M16 threaded hole and Φ 12 holes and 2.5mm grooves. These parts are matched with the brake plate, so they have high dimensional accuracy and geometric accuracy. The main content of this paper is to optimize the positioning scheme of the front support when the machining height is 2mm groove, and improve the machining accuracy of 2mm groove.

1 Introduction

With the rapid development of economic level, there are more and more categories and quantities of cars, but their prices and quality are uneven, and safety is the primary factor that people consider when buying cars. The most closely related structure to vehicle safety is the vehicle braking system.

The braking system is directly related to the life safety of every driver. The front support of a certain model studied in this subject (as shown in Figure 1, the three-dimensional model of the front support) is an integral part of the braking system and plays an important role in the braking system.

The main parts of the front support are grooved surface, 44.3mm cylindrical surface, M16 threaded hole plane, M16 threaded hole and Φ 12 holes and 2.5mm grooves. These parts cooperate with the brake plate, so they have high dimensional accuracy and geometric accuracy requirements. Their processing quality has a vital impact on the brake system.

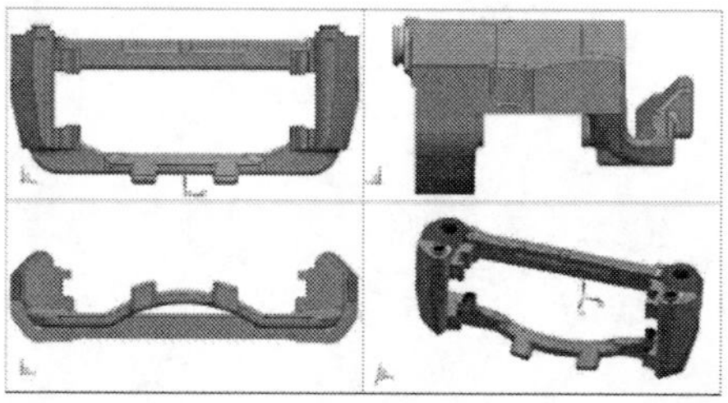

Fig.1 Three dimensional model of front support

When the existing fixture is used for processing, the 29 ± 0.25mm dimensions related to the broaching surface are out of tolerance during quality inspection. Therefore, this paper intends to analyze the positioning error of the positioning scheme.

Positioning error is the processing error caused by positioning, mainly including datum misalignment error and datum displacement error. The datum non coincidence error is caused by the non coincidence of design (process) datum and positioning datum, and the maximum position variation of process datum in the process dimension direction; The reference displacement error is caused by the inaccurate manufacturing of the positioning pair and the matching gap between the positioning pairs.See formula (1) for the calculation of positioning error:

$$\Delta D=|\Delta B\pm\Delta Y| \quad (1)$$

Among

Δ D - positioning error;

Δ B - datum misalignment error;

Δ Y - Reference displacement error.

In order to meet the processing requirements, it is generally required that the positioning error shall meet the following relationship (2):

$$\Delta D\leqslant 1/3\ T \quad (2)$$

Among T - tolerance value of process dimension.

2 Positioning scheme analysis

The main content of this paper is to optimize the positioning scheme of the front support when the machining height is 2mm groove, and improve the machining accuracy of 2mm groove. The processing part of the front support is shown in the purple area in Figure 2. The machining dimension is shown in Figure 3. The height dimension of the machining part can be obtained

* Corresponding author: 290910113@qq.com

from the main view is 29 ± 0.25, and the thickness dimension can be obtained from the left view is 27.6. According to the limit deviation value of linear dimension (GB / t1804-2000), the tolerance grade is medium, and it can be seen that the thickness dimension is 27.6 ± 0.2.

Fig. 2 View of processing parts

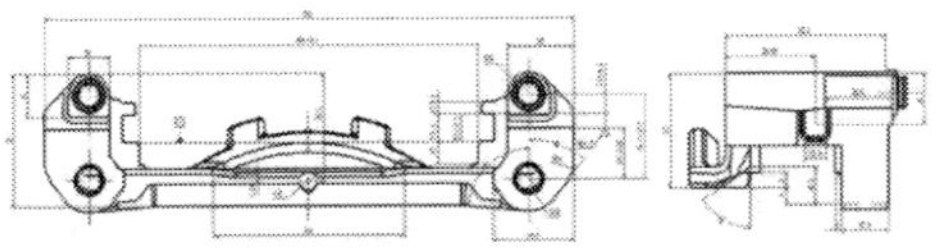

Fig. 3 Dimensions of processing parts

2.1. 29 dimensional positioning error calculation

The current positioning scheme of the front support is shown in Figure 4 ~ 5. The selected positioning datum is positioning surfaces 1, 2 and 3 in Figure 4, in which positioning surface 1 constrains 3 degrees of freedom, positioning surface 2 constrains 2 degrees of freedom and positioning surface 3 constrains 1 degree of freedom. It conforms to the six point positioning principle and belongs to complete positioning. When calculating the positioning error, it is necessary to consider the datum non coincidence error and datum displacement error. The datum non coincidence error refers to the non coincidence between the design datum and the positioning datum. It can be seen from the figure that the positioning datum does not coincide with the design datum, and the datum non coincidence error needs to be considered. Since the datum plane is a plane, the manufacturing error can be ignored in plane positioning during the analysis of datum displacement error. Therefore, the datum displacement error can be ignored during the analysis of positioning error. The design datum of the height dimension 29 ± 0.25mm related to the grooving surface is the R18 axis, while the positioning datum is the lower end face of 15 ± 0.1mm. The positioning dimension between the positioning datum and the design datum is recorded as L1. The basic dimension and its upper and lower deviation of L1 are obtained through the calculation of dimension chain (see Fig. 6 for dimension chain analysis). The calculation process is shown in equations (3) ~ (5).

$$L1=46.1-11-15=20.1 \quad (3)$$

$$EsL=+0.1-(-0.2-0.1)=+0.4 \quad (4)$$

$$EiL=-0.1-(+0.2+0.1)=-0.4 \quad (5)$$

From formulas (3) ~ (5), the basic dimension of positioning dimension L1 and its upper and lower deviation are L1 = 20.1 ± 0.4mm, which shows the positioning error Δ D is equal to the reference misalignment error ΔB is 0.8.

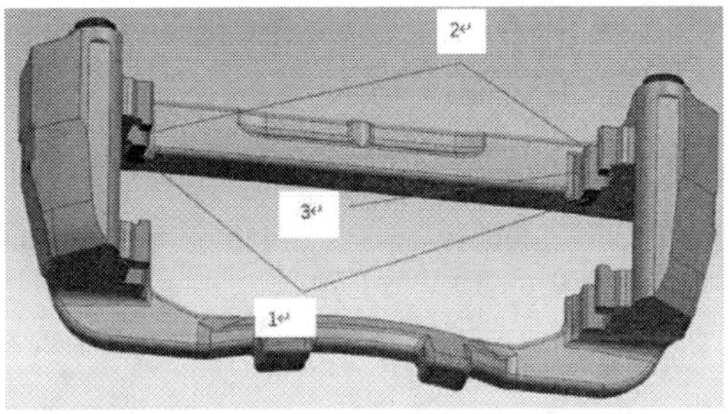

Fig. 4 3D model of positioning scheme

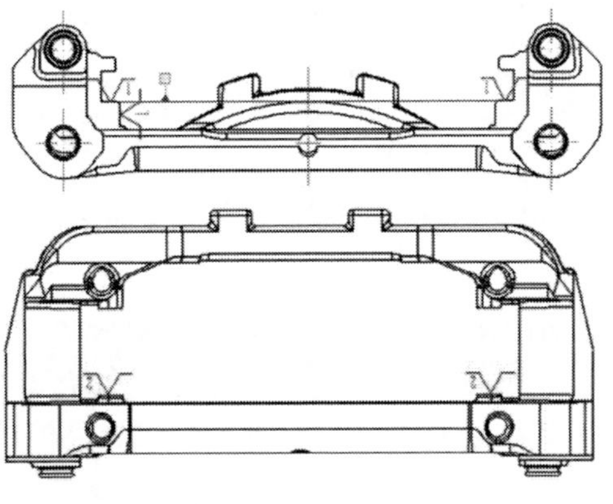

Fig. 5 Positioning scheme

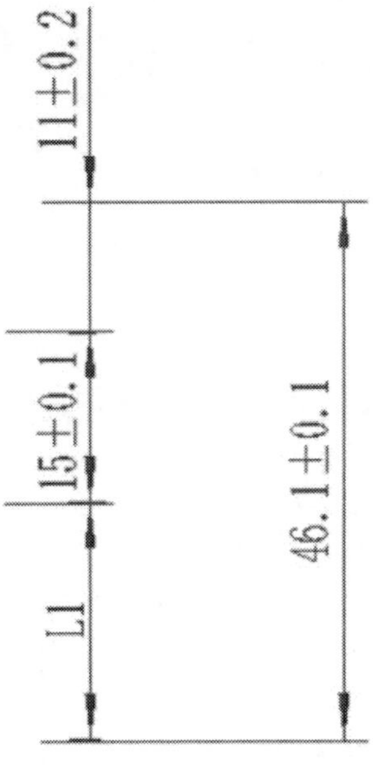

Fig. 6 Dimension chain

In order to meet the processing requirements, the calculated positioning error value shall be less than or equal to the tolerance value of 1 / 3 of the process dimension. The positioning error value of this process is

 https://doi.org/10.1051/matecconf/202135301021

0.8, and the tolerance value of height dimension 29 ± 0.25 is 0.5. Because 0.8 > 0.5, the positioning scheme of this size does not meet the processing requirements and is unreasonable, which needs to be improved.

2.2. 27.6dimensional positioning error calculation

Similarly, the design basis with thickness dimension of 27.6 ± 0.2mm is the left end face with dimension of 27.6mm in the left view, the positioning basis is the left end face with dimension of 3mm in the left view, the three-dimensional positioning surface is shown in Figure 7, and the two-dimensional positioning scheme is shown in Figure 8.

According to the limit deviation value of linear dimension (refer to national standard GB / t1804-2000), the tolerance grade is medium precision grade. According to the table, the 3mm dimension deviation is ± 0.1. The positioning dimension between the design datum and the positioning datum L2 = 3 ± 0.1mm, which shows the positioning error Δ D is equal to the reference misalignment error Δ B is 0.2, and the tolerance value of thickness dimension 27.6 ± 0.2 is about 1 / 3 of 0.4, which is equal to 0.13.

Since 0.2 < 0.13, it meets the processing requirements, and the positioning scheme in this direction is reasonable.

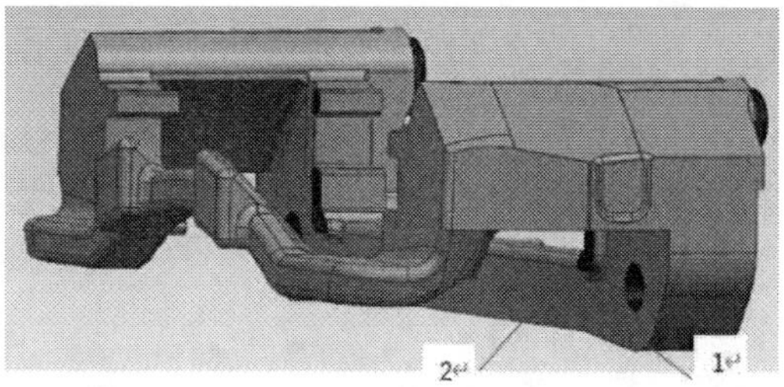

Fig.7 Three dimensional positioning scheme

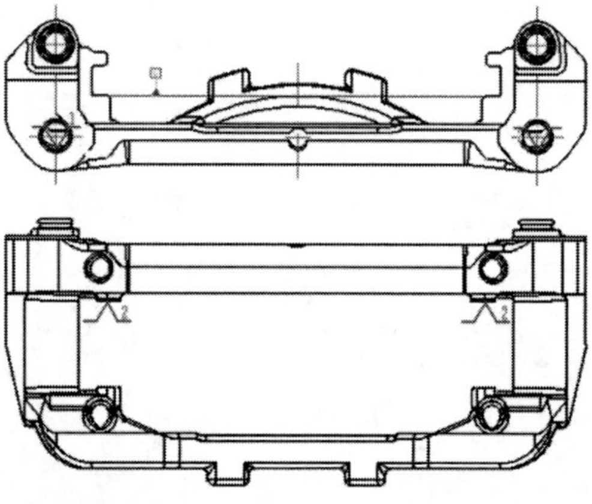

Fig. 8 Positioning scheme

3 Positioning optimization

Therefore, only the positioning in the direction of height dimension 29 ± 0.25mm needs to be improved. According to the positioning requirements of parts, the positioning in the 29mm dimension direction can be changed to 2 × ⌀ 10 hole for positioning. At this time, the positioning datum coincides with the design datum, and the positioning error in this direction is 0. The positioning scheme is shown in Figure 5-2.

Through the correction of the positioning scheme, the positioning error can be reduced and the out of tolerance problem caused by the positioning error can be avoided. According to the above calculation, when designing the workpiece positioning scheme, try to make the design datum coincide with the positioning datum, so as to reduce the influence of positioning error and improve the machining accuracy of the workpiece. At the same time, the influence of the machining accuracy of the workpiece itself and the machining accuracy of the positioning element on the positioning accuracy can not be ignored, especially in the analysis of outer circle positioning and inner hole positioning.

4 selection of positioning elements

Determine the positioning element according to the modified positioning scheme. The positioning elements used for plane positioning include support nails, support plates, adjustable supports, floating supports and auxiliary supports. The positioning elements used for inner hole positioning include cylindrical pins, conical pins, diamond pins and mandrels, among which the mandrels include cylindrical mandrels, conical mandrels and small taper mandrels. When selecting the positioning element, it should be determined in combination with the type of positioning datum plane, machining accuracy and the number of degrees of freedom to be constrained.

In the design of the improved positioning scheme, the typical positioning combination of one side and two pins is used to constrain a total of six degrees of freedom. Limited by the part structure, the positioning element for plane positioning adopts non-standard parts. The structure is shown in Figure 9, and the red surface is the limit surface of the positioning element. The inner hole is positioned by the combination of cylindrical pin and diamond pin. The structure of the locating pin is shown in Figure 10. The left side is a cylindrical pin and the right side is a diamond pin.

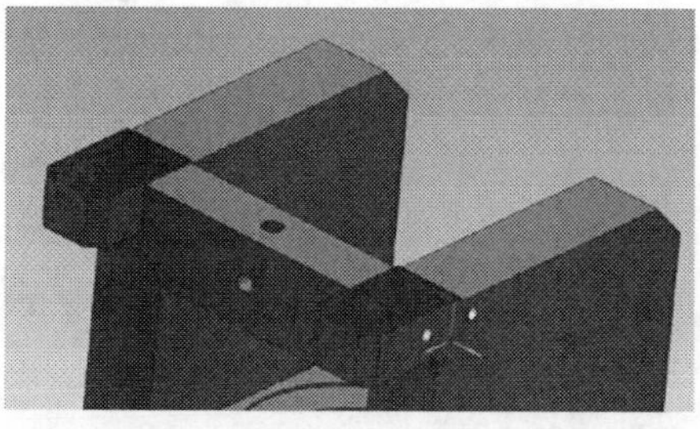

Fig. 9 Plane positioning

 https://doi.org/10.1051/matecconf/202135301021

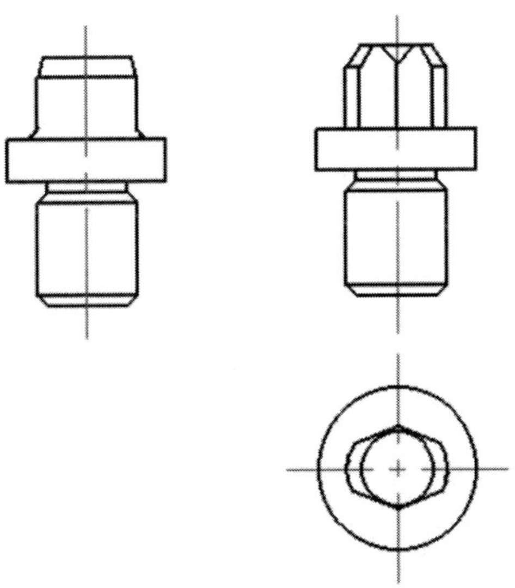

Fig. 10 Locating pin

4 Conclusion

Through the analysis of the existing positioning scheme of the front support, it is concluded that the positioning scheme in the 29mm dimension direction is unreasonable, which is also the main reason for the out of tolerance of the part in the quality inspection process. The three-sided positioning of the original positioning scheme is improved to the existing one-sided two hole positioning, so as to reduce the positioning error and make the workpiece meet the design requirements after processing. When selecting positioning elements, standard parts or non-standard parts should also be selected according to specific conditions. While considering the machining accuracy, attention should also be paid to the investment of process cost.

Acknowledgments

The research content of this paper comes from the school enterprise cooperation unit. Thank the unit for its support.

References

1. A machining-dimension-based approach to locating scheme design. Qin, Guo-Hua, Zhang, Wei-Hong,Wan, Min. Journal of Manufacturing . 2008W. Strunk Jr., E.B. White, The Elements of Style, third ed., Macmillan, New York, 1979.
2. The Design of Automobile Wheel Hub Runout Online Detection System[J] . Xiao Dong Yang,Yu Zhuo Men,Lei Yu,Jin Gang Gao. Applied Mechanics and Materials . 2015 (741-)
3. Geometric variation prediction in automotive assembling[J] . C. Xiong,Y. Rong,R.P. Koganti,M.J. Zaluzec,N. Wang. Assembly Automation . 2002 (3).
4. Systematic Modeling of Workpiece-Fixture Geometric Default and Compliance for the Prediction of Workpiece Machining Error[J]. Qin Guohua, Zhang Weihong,Wu Zhuxi,Wan Min. Journal of Manufacturing Science and Engineering. 2007 (4).
5. Computer aided fixture design: Recent research and trends[J]. Hui Wang,Yiming (Kevin) Rong,Hua Li,Price Shaun. Computer-Aided Design. 2010 (12)

MATEC Web of Conferences **353**, 01022 (2021)
AMME 2021

https://doi.org/10.1051/matecconf/202135301022

Lithium Ions Batteries Electrodes Materials, Design, Outlook and Future Perspectives

Xiangrui Li

Year 2 undergraduate, The Chinese University of Hong Kong (Shenzhen), China

Abstract. With the rapid growth of lithium-ion batteries (LIBs) industry and market, it is important to develop LIBs with higher capacities, stability and lifespan. However, challenges like low conductivity, solid electrolyte interface formation and electrode cracking now are in the way, which are closely related to electrode deficiency. Hence, to obtain better performance, the main orientation is to improve electrode materials' morphological structure. Better porosity, stability and conductivity of anode and cathode materials tangibly improve the performance of LIBs including coulombic efficiency, safety and cyclic lifespan. This review paper will combine current challenges, commonly used and emerging electrode materials together with battery testing methods and their market trend to provide a big picture about LIBs.

1 Introduction

Due to the rapid consumption of fossil fuels for modern industry and transportation together with the rapid evolution in electrical storage, it is in a pressing need to find other energy sources to replace the vacancy of fossil fuel to keep the operation of industry and catch up the market trend. The environmental concerns of greenhouse gas emission boost the development of clean energy like solar, hydroelectric and wind etc. Therefore, LIBs that are rechargeable with high energy density trait are becoming a significant direction for future energy management.

Different kinds of batteries will have different structures. Take tesla cell as an example. As shown in figure 1, this lithium-ion battery (LIB) is composed of copper, aluminium as anode and cathode electrodes. It takes graphite as anode material and lithium nickel cobalt aluminium oxide as cathode with a separator and electrolyte between them. This structure is a standard form of LIBs and its Swiss-roll like structure maximize its amount of active material for more energy[1].

ka

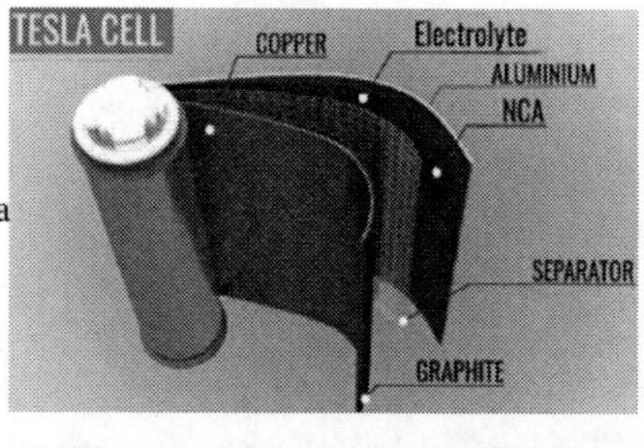

Figure 1 Lithium-ion batteries (LIBs) simplified structure

Although nowadays portable electronic devices are important consumer of LIBs, the expectation for higher capacity and large-scale equipment like electrical vehicle (EV) and grid storage also requires for high power and capacity. Recent LIBs are still insufficient in specific capacity/volumetric capacity and power density. Despite that the electric vehicles have already been commercialized at present; challenges are still evident that the modern electric vehicle nowadays can only travel for about 60-450 km according to the NEDC test cycle [2]. For the battery configuration, it takes up as large as 0.1-0.6 m^3 [3]. Besides, for limousine-class electric vehicle which has a travel range more than 300 kilometres, the weight of battery carried can even exceed 500 kg [3] (figure 2). Therefore, here comes a stringent need in assembling the next generation of LIBs with superior charge and high power that suitable for electric vehicle industry, electric appliances, and space applications [4].

Figure 2 Lithium-ion battery pack in electrical vehicle [1]

However, recent LIBs still need to be improved to match the growing demand of capacity and life cycle. The energy density of LIBs mainly depends on the mechanism of anode and cathode materials. Nanotechnology, which is an accessible science in enhancing the performance of LIBs by combining with electrode materials. Nanomaterials refers to materials that smaller than 100nm in three dimensions and cover a large span from organic to inorganic, nonmetal to metal. These nanomaterials can be customized into specified structure for different usage and have wide applications in electrode development.

2 Challenges Associated with Li-ion Batteries

2.1 Low Conductivity and Thermal Runaway

For battery, it is common that there will be capacity loss due to the low conductivity of the materials. Thermal runway is also an electrochemical problem inevitable caused by the internal impedance of the battery. Thermal runaway is a major concern for LIBs application [2], and this problem is commonly occurred in the lithium transition metal oxide, for example, lithium cobalt oxide has the lowest thermal stability of any commercial cathode material [37]. This problem will cause the energy runoff in the form of heat when operating. Heat dissipated from the LIBs is mainly generated by activation (interfacial kinetics), concentration (related to the transportation of species) and ohmic losses (resistance) during the electron movement [4]. To reduce the low conductivity and thermal runaway, electrode and electrolyte materials with low impedance should be exploited. Commonly used method is to incorporate the carbon or highly conductive agents at the external surface or the inner surface of electrode [5]. All these issues related to the electrochemical property of materials and it's important to figure out their chemical behaviours to control the thermal runaway.

2.2 Morphological and Structural/Phase Changes

Morphological structure is a vital characteristic of electrode materials and it is a key concern in the LIBs since it relates to the process of Li insertion and extraction, which can primarily affect the battery performance. Moreover, morphological variations will accumulate isolated products that lead to the electrical contact loss that obliterates the cell operation [4]. Not only morphological variation, materials will also display crystal structure and phase deviation that affect the battery capacity and lifespan [4]. Change in crystal structure will cause defective conductivity and inactive kinetics for Li+ insertion and extraction process, incurring poor capacities and lifespan [6]. Therefore, to solve this problem, scientists involve nanoscience in to refine materials' nanostructure that can partially diminish the morphological deviations by reinforced phase conversion kinetics [4].

2.3 Volume Changes and Thickness of Solid Electrolyte Interface Film

Volume change happens during Li+ alloying/dealloying process because Li+ will apply strains to the electrodes. Solid Electrolyte Interface (SEI) is a layered substance that usually forms on the surface of anode and cathode due to the electrochemical reduction of the electrolyte and usually occurs during the initial few charge-discharge cycles of LIBs [32]. The formation of the SEI is connected with the electrochemical properties of the electrolyte and the electrode materials and will affect the performance of battery including its capacity, safety and life cycle [2, 4, 32]. During the charging/discharging process, the movement of particles (ions and electrons) leads to the redox reaction between Li-ions, electrons and electrolyte (mostly lithium salts like $LiPF_6$ and organic solution) (Figure 3a) and form the SEI layer. As for unsustainable SEI, it cannot sustain stress caused by the volume change and easily breakdown. This breakdown will further lead to the exposure of electrode and SEI formation (Figure 3b), but it can be predicted by the morphological and volumetric variations for proactive control [4]. Simultaneously, it can also serve as a protective coating of the electrode that allows lithium ions to pass for intercalation but not electrons [42]. Therefore, it can somehow prevent the further degradation of the electrolyte. Under this condition, it is important to control the thickness of SEI to maximize the battery performance. However, up to now, there are no preferable mechanisms to restrain the SEI layer compositions, grain size, thickness and spatial distribution [4]. Hence, morphology of electrode materials affects the formation of SEI film predominantly. Customized electrode morphology and encasing the appearance with inactive materials like carbon coatings will improve our control over SEI film [4].

2.4 Cracking and Fracture of Particles and Electrodes and Electron/ion Transport

Cracking of particles and electrodes commonly occurs during the lithium intercalation/ extraction, resulting the further corrosion of electrode from electrolyte through these cracks. During the intercalation and deintercalation process, the frequent egress and ingress of Li+ will apply strains on electrode materials and ultimately cause the cracks. Generally, traditional intercalation-type electrode materials undertake subtle volume variations smaller than 10%. On the contrary, novel electrode materials with high capacity suffer greater volume variations because of large lithium intake [4]. The large variation in the volume during lithium intercalation/extraction process is considered as a crucial issue in restricting the performances and applications of high-capacity electrodes.

Alloy materials are generally characterized with high capacity due to large lithium intake and therefore will undertake great volume variation, for example, 420% for Si, 300% for P, and 260% for Ge and Sn [4]. These drastic volume changes lead to the electrodes, active materials potential, and the cycle life degradation [4]. To solve the problems of cracking and facture, commonly used method is to coat the electrodes with materials that nonreactive to the electrolyte to prevent the direct contact between electrode and electrolyte, and then restrain the erosion from electrolyte. These materials should be permeable for lithium ions but hinder the electron movement to restrict the formation of SEI.

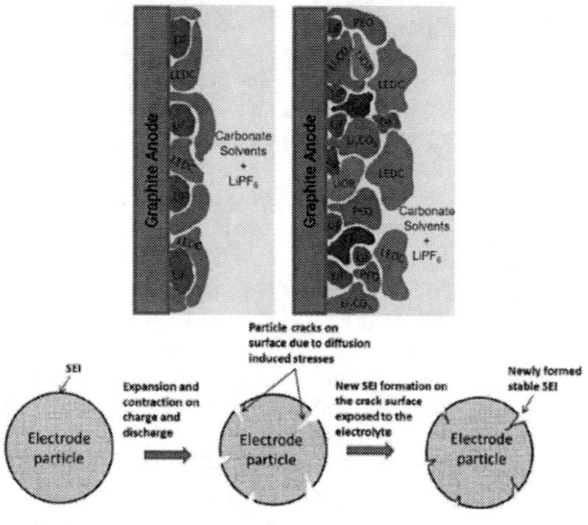

Figure 3 (a) SEI composition and distribution [7]; (b) Electrode fracture and SEI formation process [8]

3 Anode Materials

3.1 Carbon-Based Electroactive Materials

Carbon-based materials include carbon nanotubes (CNTs), graphene, fullerenes and other materials that composed of carbon atoms arranging in different morphological structure (Figure 4). Commonly used materials for anode electrode are CNTs, graphene who have excellent electron conductivity, stability and transportation ability for lithium intercalation [4, 43]. Besides, carbon-based materials with attractive benefits of relatively low cost, moderate energy density, power density and electrochemical stabilities, who are suitable for LIB anode materials [2, 4].

Conventionally, graphite is one of the most commonly used 2D carbon-based anode materials. It is composed of sp2 carbon atoms with honeycomb-like structure [4]. And it derivative – graphene, has drawn great attention due to its impressive properties and extensive applications in many fields, including physical, chemical, and biological [9]. Its special latticed single layer morphological structure contributes to excellent electron conductivity, mechanical strength, well mobility and great specific surface, making it possible to be a competitive candidate in LIB applications [10].

As for the CNTs, it is a universal 1D nanomaterials used in fabricating porous body and combining with other anodic materials to improve electrode performance due to its excellent electronic conductivity, thermal, mechanical stabilities and good transportation rate [2, 4, 43]. CNTs have various structures. It can be roughly categorized into single-walled carbon nanotubes (SWCNTs) and multi-walled nanotubes (MWCNTs) [4]. It also has different grid structures including armchair, zigzag and chiral, where armchair is conductive, zigzag is semiconductive and chiral is insulative.

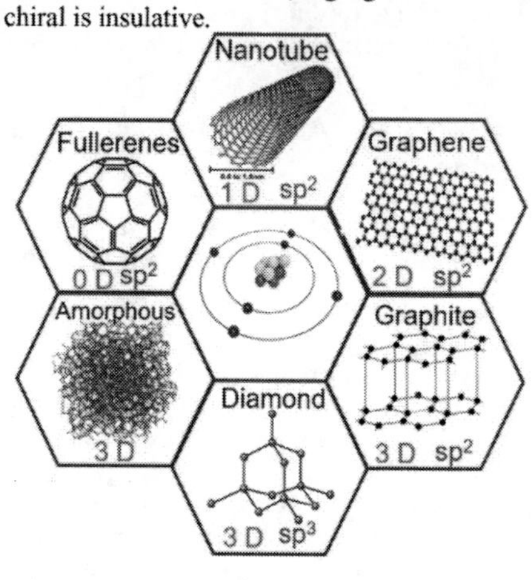

Figure 4 Structures of different kinds of carbon-based materials [12]

3.2 Titanium-Based Oxides Materials

$Li_4Ti_5O_{12}$ is one of the memebr that has been successfully commercialized because it excellently combines the factors of outstanding thermal stability [13], relatively high volumetric capacity, and long-lasting lifespan, inspite of flaws like inferior cell voltage, high cost and low capacity (175 m Ah g^{-1} & 600 m Ah cm^{-3}) [2]. It was also considered "zero strain" under high potential lithiation/delithiation because the phase change is tiny (0.2%) in volume [2]. Therefore, this slight volume change can increase the stability of both electrode and SEI film. Furthermore, LTO can offer high potential above 1V that can avoid the formation and growth of SEI film. The high potential also prevents the Li dendrite formation which will cause capacity loss and bad safety [2].

Due to the property of the titanium-based oxide, the gassing reaction on the surface is inevitable between organic electrolyte and LTO [38], which will cause the capacity loss. This process can be suppressed by coating carbon materials but the involving of carbon materials also lead to problems of catalyzing electrolyte decomposition [14, 15]. But in general, LTO anode can last for tens of thousands cycles and become a competitive material in high power application [39].

3.3 Alloy materials (Si, Gr, Metal oxides, Metal Phosphides)

Alloy materials are characterized with high capacity. Germanium, silicon, silicon oxide and tin oxide are anodic materials that have high capacity taking place in lithium reaction correlated to alloy/de-alloy phenomenon [4]. Their specific capacity can range from 783 mA h g^{-1} for tin oxide to 4211 mA h g^{-1} for silicon [16-18]. However, this trait also leads to the electrode expansion and further cause short lifespan. There are two solutions regarding to this problem: narrowing the particles size to nanoscale and manufacturing compositions with the combination of active lithium and inert materials [4]. With nanoscale particles, the strain given to the electrode will be decrease and protect the electrode. For the second method, active/inactive lithium material is more receptive by this composite because it can be regarded as a buffer between the inert materials and active lithium [19], which makes intercalation easier.

Germanium is a potential anode material with relative high capacity (1623 mA h g^{-1}) compared to common anodes [4]. But it has drawbacks of higher cost and lower capacity in comparision with silicon, it is competitive in the fields of higher electronic conductivity that 104 times greater than silicon and narrow bandwidth potential (0.67 eV) [4]. Even so, the practical application of germanium is hampered due to its large volume variation (300%) [4]. That's why nanoarchitectures are introduced to eliminate this kind of problem. Nanotubes, nanowires and nanoparticles [20, 21, 22] can conspicuously constraint the volume change more efficiently when compared to bulk and microstructure. These nanoarchitectures can work as a structural buffer to restrict the volume variation and can even reduce the formation of SEI layer [4].

Silicon is an emerging anodic material with extremely high capacity (e.g : $Li_{22}Si_5$, 4200 mA h g^{-1}) [4]. It is also abundant on the earth with relatively low price compared to lithium. Si is an outstanding candidate due to its relatively low delithiation potential, high gravimetric/ volumetric capacity, chemical stability, reserves, and non-toxicity [2]. Like other alloy materials, it will suffer great variation in volume. To solve this problem, silicon can be combined to MXene that form interspaces to orderly receive Li-ions and prevent expansion (figure 5).

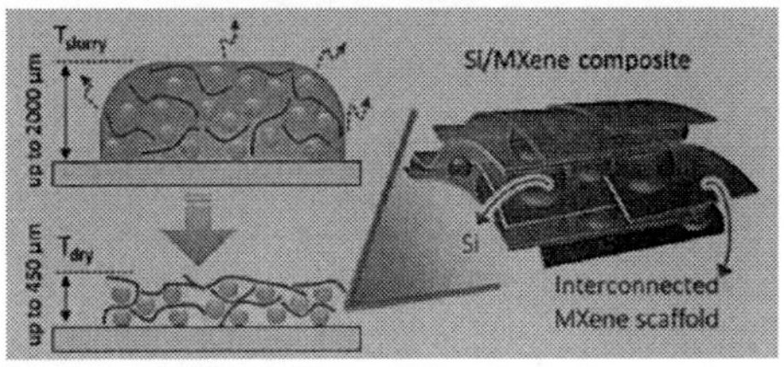

Figure 5 Slurry-casting process of Si/MXene combination and its structure [4]

Silicon oxide has capacity over 1600 mA h g^{-1}, lower volume charge and simultaneously low activation of energy [4]. SiO also represents a dominating disadvantage: it is not thermally stable at various temperatures [4] and will be converted to SiO_2 at high temperature ambience, which will damage the capacity and lifetime of the battery. The involvement of oxygen atom will lead to the formation of Si-Li alloy and also the lithium silicates [4] during the discharge process, and this can be described in the following equation:

$$\mathrm{xLi} + \mathrm{SiO} \leftrightarrow \mathrm{Li_x} - \mathrm{ySi_z} + \mathrm{Li_xSiO_z} \quad (1)$$

4 Cathode Materials

4.1 Lithium Cobalt Oxide (LiCoO2)

Lithium cobalt oxide is a widely used cathodic material in LIBs. $LiCoO_2$ (LCO) can provide layer structure for lithiation and the lithium cations lie in between two layers of composition of cobalt and oxygen atoms like sandwich. LCO is a fantasting cathode material with high specific capacity (274 mA h g^{-1}), high volumetric capacity (1363 mA h cm^{-3}), mild idle self-discharge, high discharge voltage, and good cycling performance [40, 41]. The major limitations of it are low thermal stability, and fast capacity fade at high current or deep cycling [2]. This low thermal stability is reflected in high temperature performance, which will cause exothermic oxygen release and lead to flame [2].

4.2 Lithium Nickel Oxide (LiNiO2)

Lithium nickel oxide is characterized by high capacity which is widely used in portable devices and it can offer specific capacity for about 200m Ah g^{-1} [4]. Compared with LCO, it has similar structure to allow the lithium-ion intercalation but harder to produce because the trivalent nickel is unstable in high temperature environment [4]. And LNO's cyclic stability is poor, which is ascribed to the cation mixing [44]. Also, LNO is unstable at high temperature environment because the +3 nickel is unstable at that situation [4]. Therefore, there is method called doping to improve its drawbacks by involving other particles [4].

4.3 Lithium Manganese Oxide (LiMnO2)

Lithium manganese oxide has many different structures and $LiMnO_2$ is one of them that has layered structure for lithium intercalation and suitable for cathode material. $LiMnO_2$ (LMO) is a commonly used cathode in primary cell which can also be used in rechargeable cell thanks to its low cost, envrionmental implication, and cost-effective alternative to LCO and LNO [4]. But its cyclic performance is not satisfied because its layered structure is not stable during the extraction and Mn will run off during the charge-discharge process [2], providing a bad cyclic stability.

4.4 Combination

The combination cathode refers to the combination of cobalt, nickel and manganese in the form of Li $[Ni_{1-x-y}Co_x Al_y] O_2$ to acquire the advantages of above transition metals, where nickel can provide high specific capacity, cobalt can offer enhanced rate capability and manganese can enhance structural integrity with its layered structure together with better safety in battery [4, 23]. Therefore, the proportion of each element in the cathode determine the performance of the final product. Recently, there are researches about the high nickel LIB, like NCM631, NCM 811 (numbers after stands for the ratio of different components). Although it has some outstanding point like high capacity, but high nickel content simultaneously means that the cobalt and manganese contents will be lower, which will cause the overall structural instability and serious capacity fading [25], also with safety issues [4]. High nickel content will also form spinel at the surface, the inactive phase Fm (-) 3m and the fatigue capacity that damage its performance [4] [26].

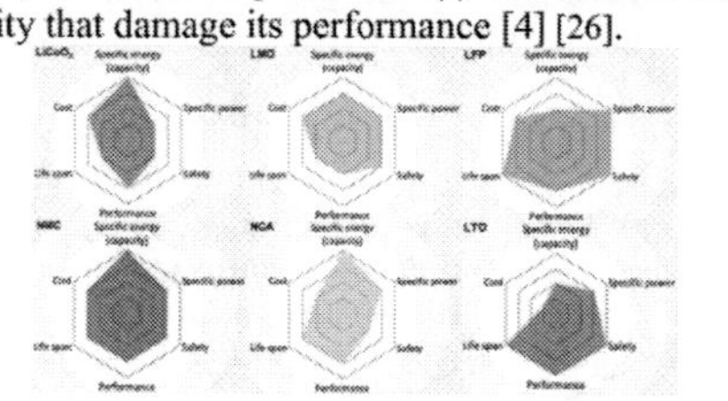

Figure 6 Comparisons of different LIBs in the perspectives of specific energy, power, safety, performance, lifespan, and cost.

4.5 Iron Fluoride

Generally, the mono electron diffusion involved during the intercalation process limited the capacity of LIBs [27]. Iron trifluoride has become an incaparably important material because of its eminent features of low cost, non-toxity and satisfactory thermal stabilities [28]. And the emerging material FeF_3 can provide eminent capacity, energy density and great voltage output because of its high electronegativity from both iron and fluoride [4]. The manufacturing process of FeF_3 so far mostly involves in solution which means that the employment of HF is unavoidable [28]. The HF has strong corrosivity, toxicity so that it is dangerous for mass production and hinder its commercialization.

4.6 Vanadium Oxide

Among aforementioned cathode materials, vanadium pentoxide (V_2O_5) and other vanadium-based oxides have their attractive features including enhanced specific capacity, low cost, facile synthesis and abundance which stand out and comparable for LIB cathode material [4]. Vanadium pentoxide has orthorhombic structure and interconnect each molecular by sharing the oxygen atom, then forms layered structure and each layer can be restrained by an der Waals forces. This layered structure offers good capacity for Li^+ intercalation and extraction, resulting good synthesized capacity value up to 294 mA h g^{-1} [29]. Moreover, vanadium is relative common and has lower price compared to lithium (Figure 6).
However, the practical use of V_2O_5 is restricted according to its low Li+ diffusion rate (10^{-12}–10^{-13} cm^2 s^{-1}), restricted conductivity (10^{-2}–10^{-3} S cm^{-1}) and unstable structure. All these factors will slow the ions and electron transportation and lead to poor cyclic stability and overall performance [30] [31].

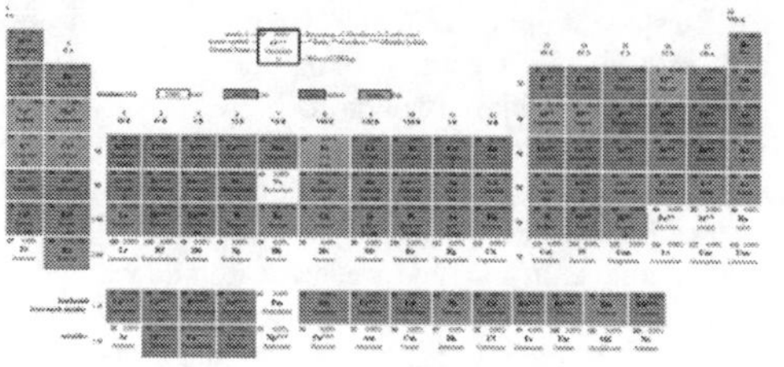

Figure 7 Price and abundance of elements in periodic table

5 Li-ion battery Testing

5.1 Galvanostatic charge-discharge cycling behaviour

The galvanostatic charging and discharging behaviors are important factors to obtain the voltage, charging or discharging behaviors and columbic efficiency (CE). The galvanostatic charge-discharge curve (GCD) is a commonly used chart for analyzing battery cycling behavior. As shown in figure 8a, the GCD curve always consist of an ascending line and a descending line which represent the charging and discharging process relatively. From the curve, the voltage behavior can be observed according to the y-axis. Furthermore, the charging capacity of the battery can be obtained if we calculate the integral before the summit point. Likewise, the discharging capacity can be obtained and then the CE can be calculated by the formula:

$$CE = \frac{Q_c}{Q_d} = \frac{A_b}{A_a} \quad (2)$$

where Q_c, Q_d represents the charge capacity and discharge capacity, A_b represents area before summit and A_a stands for area after summits. Generally, the symmetric curve means the 100% CE but a concave curve connected to a convex curve represents CE < 100%. The GCD curve also can present charge/discharge curves separately with state of charge (SoC) x-axis (Figure 8b) or specific capacity, all these depend on the requirement of experiment.

5.2 Cyclic stability

For all kinds of batteries, their performance will fluctuate due to its internal variation after serval cycles. Therefore, to test and improve battery stability, cyclic stability is an important indicator to be measured. Shown here in 8c, it represents the specific capacity and coulombic efficiency of different materials at the same time and it can also take volumetric capacity as y-axis as well. Generally, the capacity retention of the battery will decrease due to lithium loss and other factors and then tend to be stable. However, it can be observed that the coulombic efficiency will have a sharp increase at the beginning oppositely then become stable. The reason is that for a fresh battery, it takes time for SEI layer to form during the innitial several cycles[32]. From this diagram, capacity and CE variation and their steady state are generally important target that should be paid attention to.

5.3 Internal resistance

Open circuit voltage is a primary characteristic of a battery. Open circuit voltage is the voltage between the positive and negative terminal of the battery when there is no discharging current or loads. The value of the open circuit voltage is determined by the redox reaction of the active materials and it can be directly measured by the voltameter or multimeter directly. For practical use, the battery cannot reach to open circuit voltage because of its internal resistance when current applied. The internal resistance of battery will take away some voltage. With the voltage take away by battery and current through it, its internal resistance can be sketchily calculated by

$$R = \frac{V_b}{I_b} \tag{3}$$

where V_b represents voltage taken by internal resistance and I_b represents the current through it. Hence, when fabricating the battery, the pursuit of low internal resistance is always important in reducing the thermal runaway and energy loss.

5.4 Nyquist plots

Nyquist plot is the diagram to obtain material's impedance. Firstly, it is important to know the concept of impedance. Impedance refers to the ability of the electrolyte that prevents the ions to move freely to the electrode and related to the ion diffusion rate. This resistance includes the real part resistance like a resistor and also imaginary part resistance always contributed by a capacitor. Generally, the real part is denoted as R and the imaginary part is presented in the imaginary number j (j = sqrt (-1)). And Electrical impedance spectroscopy (EIS) is the method that used to research on the impedance of the battery and analysis its diffusion coefficient. The Nyquist plot is the direct demonstration of EIS in different frequencies.

Like figure.1, the Nyquist plot is set up in a coordinate system of x-axis (real part) and y-axis (imaginary part) with function Z(w), where Z represents the impedance and w represents the frequency. The plots start with high frequency of a semicircle to low frequency of a sloping line, where semicircle represents the reaction process and sloping line refers to diffusion process respectively [45]. And the first intersection point refers to the solution resistance and the second intersection point represents charge transfer resistance [45]. The impedance caused by diffusion process is also called Warburg impedance which is related to diffusion rate. Therefore, data from the Nyquist plot can be used to compute the diffusion coefficient.

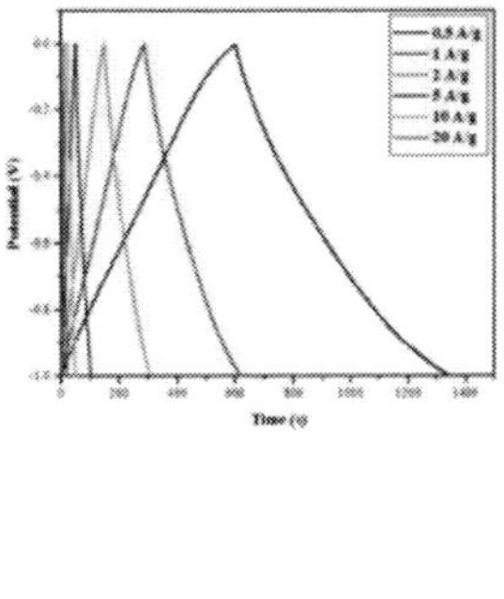

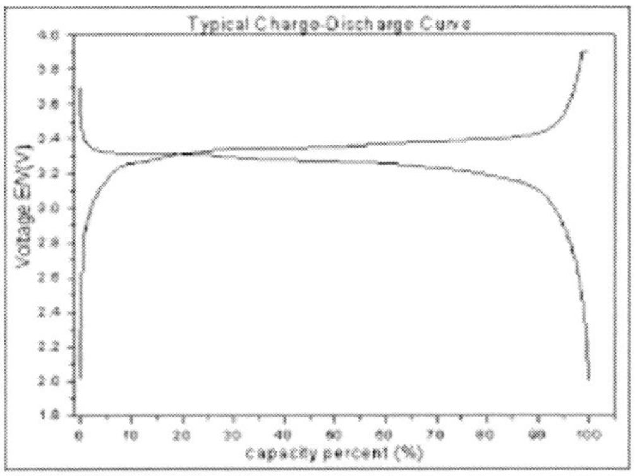

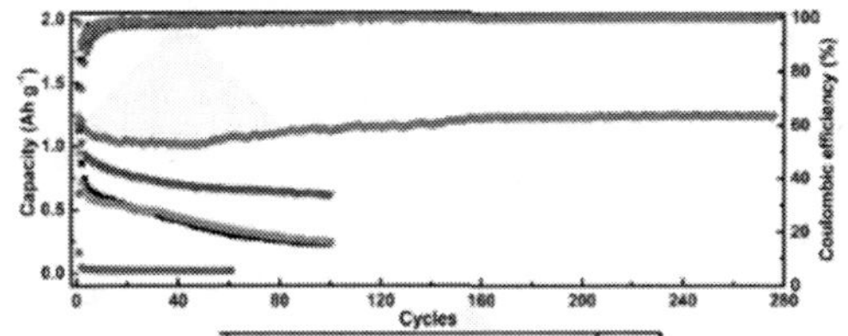

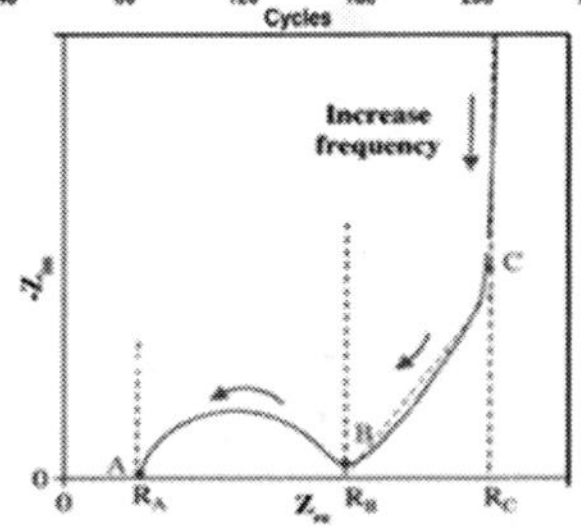

Figure 8 (a) Galvanostatic charge-discharge (GCD) curve with different specific current [6]; (b) Charge-discharge curve [33] (c)charge-discharge curves that represent the capacity and coulombic efficiency variation of different materials (represented by different colors) [34] (d) Nyquist plots. The red pointer points from high frequency to low frequency [35]

6 Market Trends

Nowadays, there exist a large number of companies dealing with nanomaterials and battery fabrication because of the growing demand from fields like electronic devices, transportation and grid storage.

As table 4 shows, portable devices including mobile phones, tablets, laptops and other electronic devices had become the primary LIBs market which constitute about 80% of the share at the time of 2015 and is still increasing steadily. The cosumers demand varied over years, for example, from the table, it can be seen that the annual LIB demand for laptops (PC) is relatively stable at about 10 GWh, while more portable devices are being preferred and gowing firmly. The LIB demand for cell phones and tablets is growing at an average speed of 10% annually. From 2015 to 2020 the average annual growth rate of portable devices is 9%, and 10% from 2025 in 2030.

For the transportation, LIBs are mostly used in electrical vehicle (EV), hybrid electric vehicle (HEV) and plug-in hybrid electric vehicle (PHEV). The market trend is exceeding the portable electronic field because of the growing of new energy vehicle industry. In 2025, the

https://doi.org/10.1051/matecconf/202135301022

estimated demand for road transportation will reach to 137GWh and 245GWh in 2030. The average annual growth rate from 2025 to 2030 is 16%, which is considered to take the place of the leading position of portable devices.

Also, the emergence of energy storage is a latent consumer of LIBs in the future. LIBs can be utilized as storage units in power supply systems due to its high capacity and modularization trait for both grid-connected (centralized) and off-grid (distributed) [36]. Until 2015 the share of storage is still 0 GWh but grow in a rapid rate in the following years, the estimated annual growth rate from to 2030 even reaches 40%, which is a competitive market of LIBs in the future.

Table 1 Li-ion battery market, previous development and future prospect from 2010 to 2030 [36]

	2010	2015	2020	2025	2030
Cell phones	6	11	17	28	44
Tablets	1	7	12	17	25
PC	12	9	9	9	11
Portable electronics, other	3	4	7	12	20
Portable electronics, total	**21**	**31**	**45**	**66**	**100**
EV	0	11	65	115	200
PHEV	0	2	8	13	25
HEV	0	0	2	7	15
Road-transport, other	0	0	1	2	5
Road-transport, total	**0**	**13**	**76**	**137**	**245**
Storage in power supply	**0**	**0**	**2**	**10**	**30**
Other applications	**1**	**1**	**2**	**7**	**15**
Total	22	45	125	220	390

7 Conclusion and Future Perspectives

The aforementioned article has already presented a framework for LIBs current challenges, electrode materials, testing method and their market trend. LIB is an indispensable source in the future and nanomaterials are significant way to enhance its capability .

In the future, we are looking forward to batteries with better property (Figure 9), including better capacity, lower price, safety etc. The challenges nowadays are low conductivity, electrode fracture and SEI layer which are tightly related to the electrode physicochemical property and morphological structure. When considering about suitable electrode materials, basically their cost, conductivity, chemical property and morphological structure are always need to comprehensively considered. Not only materials themselves, the combination of different compositions can also auxiliary improve their abilities or shortcomings. Normally, the combination of nanomaterials including, graphene, CNT and other nanoparticles with electrode materials is a significant way in developing better electrode with excellent stability, conductivity and capacity.

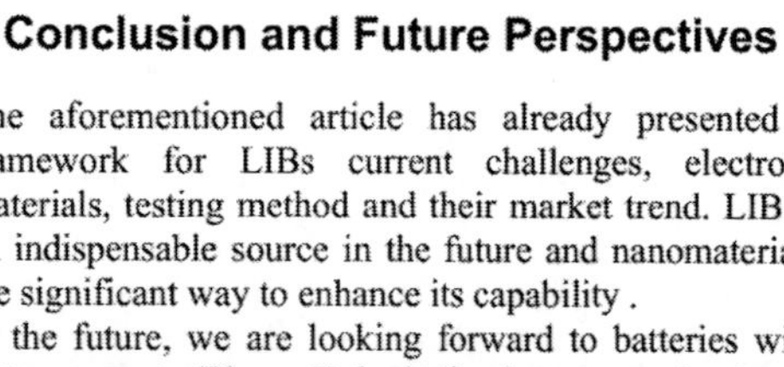

Figure 9 Significant improvement needed in the future [12].

Better capacity is always what we pursue, and researcher are trying hard to develop materials that can be apply extensively. For cathode materials, the combination electrodes especially nickel-rich cathode is developing rapidly with excellent capacity. Currently, a variety of NCM members like NCM333, NCM523 are already commercialized and the enhanced materials NCM811 and NCM622 are still under research and hopefully they can be deployed in the near future [4]. The testing method including GCD curve, Cyclic stability analysis and Nyquist plot are important in analyzing battery capacity, stability and conductivity in a visualized and quantified way. Hopefully, the battery market trend will definitely continue to rise, especially portable devices, electrical vehicle (EV)) and grid storage. Hence, it is important to match the steeply growing demand and various requirement of different equipment.

Even though electrode material is an important and potential factor to develop battery performance in the following years, there still other parts of battery like separator, electrolyte and internal structure can contribute to better LIBs which needs to be further studied.

References

1. Lithium-ion battery, how does it work? https://www.youtube.com/watch?v=VxMM4g2Sk8U
2. Nitta N, Wu F, Lee J T, et al. Li-ion battery materials: present and future. Materials today, 2015, 18(5): 252-264.
3. Thomas C.E. Fuel cell and battery electric vehicles compared. International Journal of Hydrogen Energy. 2009, 34 (15), 6005-6020.
4. Kim H J, Krishna T N V, Zeb K, et al. A comprehensive review of Li-ion battery materials and their recycling techniques. Electronics, 2020, 9(7): 1161.
5. Bai, P.; Bazant, M.Z. Charge transfer kinetics at the solid-solid interface in porous electrodes. Nat. Commun. 2014, 5, 1–7.
6. Son, M.Y.; Hong, Y.J.; Lee, J.K.; Chan Kang, Y. One-pot synthesis of Fe2O3 yolk-shell particles with two, three, and four shells for application as an anode material in lithium-ion batteries. Nanoscale 2013, 5, 11592–11597.
7. Heiskanen S K, Kim J, Lucht B L. Generation and evolution of the solid electrolyte interphase of lithium-ion batteries. Joule, 2019, 3(10): 2322-2333.
8. Xu, J; Deshpande, Rutooj; Pan, J; Cheng, Y; Battaglia, V. Electrode Side Reactions, Capacity Loss and Mechanical Degradation in Lithium-Ion Batteries. Journal of The Electrochemical Society. 2015, 162. 2026-2035.
9. Stankovich, S.; Dikin, D.A.; Dommett, G.H.B. et al. Graphene-based composite materials. Nature 2006, 442, 282–286.

10. Liang, M.; Zhi, L. Graphene-based electrode materials for rechargeable lithium batteries. J. Mater. Chem. 2009, 19, 5871–5878.
11. Brownson, D.A.C.; Kampouris, D.K.; Banks, C.E. An overview of graphene in energy production and storage applications. J. Power Sources 2011, 196, 4873–4885.
12. M. B. Bazbouz. Lithium-ion Battery Materials (Anode, Cathode and separator). 2021.
13. Doughty D H, Roth E P. A general discussion of Li ion battery safety. The Electrochemical Society Interface, 2012, 21(2): 37.
14. He Y B, Liu M, Huang Z D, et al. Effect of solid electrolyte interface (SEI) film on cyclic performance of Li4Ti5O12 anodes for Li ion batteries. Journal of Power Sources, 2013, 239: 269-276. 15. M.-S. Song, et al. J. Mater. Chem. 2014, A2(3), 631.
15. Reddy, A.L.M.; Gowda, S.R.; Shaijumon, M.M.; Ajayan, P.M. Hybrid nanostructures for energy storage applications. Adv. Mater. 2012, 24, 5045–5064.
16. Mekonnen, Y.; Sundararajan, A.; Sarwat, A.I. A review of cathode and anode materials for lithium-ion batteries. In Proceedings of the SoutheastCon 2016, Norfolk, VA, USA, 30 March–3 April 2016.
17. Kong, L.L.; Wang, L.; Ni, Z.C. et al. Lithium–Magnesium Alloy as a Stable Anode for Lithium–Sulfur Battery. Adv. Funct. Mater. 2019, 29, 1808756.
18. Kasavajjula, U.; Wang, C.; Appleby, A.J. Nano- and bulk-silicon-based insertion anodes for lithium-ion secondary cells. J. Power Sources 2007, 29, 1808756.
19. Park, M.H.; Cho, Y.; Kim, K. et al. Germanium nanotubes prepared by using the Kirkendall effect as anodes for high-rate lithium batteries. Angew. Chem. Int. Ed. 2011, 50, 9647–9650.
20. Flynn, G.; Palaniappan, K.; Sheehan, M.; Kennedy, T.; Ryan, K.M. Solution synthesis of lead seeded germanium nanowires and branched nanowire networks and their application as Li-ion battery anodes. Nanotechnology 2017, 28(25): 255603.
21. Kim, C.H.; Im, H.S.; Cho, Y.J. et al. High-yield gas-phase laser photolysis synthesis of germanium nanocrystals for high-performance photodetectors and lithium-ion batteries. J. Phys. Chem. C 2012, 50, 26190–26196.
22. Scrosati, B. History of lithium batteries. J. Solid State Electrochem. 2011, 15, 1623–1630.
23. Blomgren, G.E. The Development and Future of Lithium-Ion Batteries. J. Electrochem.Soc. 2017, 164, 1.
24. Hwang J, Do K, Ahn H. Highly conductive 3D structural carbon network-encapsulated Ni-rich LiNi0.8Co0.1Mn0.1O2 as depolarized and passivated cathode for lithium-ion batteries. Chemical Engineering Journal, 2021, 406: 126813
25. Is Li-Ion the Solution for the Electric Vehicle? https://batteryuniversity.com/learn/archive/is_li_ion_the_solution_for_the_electric_vehicle
26. Zheng, Q.; Yamada, Y.; Shang, R. et al. A cyclic phosphate-based battery electrolyte for high voltage and safe operation. Nat. Energy 2020, 5, 291–298.
27. Li, T.; Li, L.; Cao, Y.L. et al. Reversible three-electron redox behaviors of FeF3 nanocrystals as high-capacity cathode-active materials for Li-Ion batteries. J. Phys. Chem. C 2010, 114, 3190–3195.
28. Chen, C.; Xu, X.; Chen, S. et al. The preparation and characterization of iron fluorides polymorphs FeF3·0.33H2O and β-FeF3·3H2O as cathode materials for lithium-ion batteries. Mater. Res. Bull. 2015, 64, 187–193.
29. Ma, Y.; Huang, A.; Zhou, H.; Ji, S.; Zhang, S.; Li, R.; Yao, H.; Cao, X.; Jin, P. Template-free formation of various V2O5 hierarchical structures as cathode materials for lithium-ion batteries. J. Mater. Chem. 2017, 5, 6522–6531.
30. Lai Q, Sun Y, Wang T, et al. How to design hydrogen storage materials? Fundamentals, synthesis, and storage tanks. Advanced Sustainable Systems, 2019, 3(9): 1900043.
31. Smith A, Burns J, Trussler S, Dahn J. Precision measurements of the coulombic efficiency of lithium-ion batteries and of electrode materials for lithium-ion batteries. J Electrochem Soc 2010, 157(2): A196-202
32. Qaisar, S M; Dallet, D; Benjamin, S; et al. Power efficient analog to digital conversion for the Li-ion battery voltage monitoring and measurement. IEEE International Instrumentation and Measurement Technology Conference (I2MTC). IEEE, 2013: 1522-1525.
33. Shah, M S A S; Park, A R; Rauf, A et al. Highly interdigitated and porous architected ternary composite of SnS2, gC3N4, and reduced graphene oxide (rGO) as high-performance lithium-ion battery anodes. RSC Advances, 2017, 7(6): 3125-3135.
34. Mei, B., Munteshari, O., Lau, J., Dunn, B., & Pilon, L. Physical Interpretations of Nyquist Plots for EDLC Electrodes and Devices. Journal of Physical Chemistry C, 2018, 122: 194-206.
35. Zubi G, Dufo-López R, Carvalho M, et al. The lithium-ion battery: State of the art and future perspectives. Renewable and Sustainable Energy Reviews, 2018, 89, 292-308.
36. Doughty D H, Roth E P. A general discussion of Li ion battery safety. The Electrochemical Society Interface, 2012, 21(2): 37.
37. He Y B, Li B, Liu M, et al. Gassing in Li4Ti5O12-based batteries and its remedy. Scientific reports, 2012, 2(1): 1-9.
38. Zaghib K, Dontigny M, Guerfi A, et al. Safe and fast-charging Li-ion battery with long shelf life for power applications. Journal of power sources, 2011, 196(8): 3949-3954.
39. R. Yazami, et al. New Trends in Intercalation Compounds for Energy Storage and Conversion:

Proceedings of the International Symposium. The Electrochemical Society, vol. 2003, 2003, p. 317.

40. Du Pasquier A, Plitz I, Menocal S, et al. A comparative study of Li-ion battery, supercapacitor and nonaqueous asymmetric hybrid devices for automotive applications. Journal of power sources, 2003, 115(1): 171-178.
41. Liu, N., Lu, Z., Zhao, J. et al. A pomegranate-inspired nanoscale design for large-volume-change lithium battery anodes. 2014, Nature Nanotech 9, 187–192.
42. Yu, Y.; Cui, C.; Qian, W. et al. Carbon nanotube production and application in energy storage. Asia-Pac. J. Chem. Eng. 2013, 8, 234–245.
43. Ohzuku, T. Electrochemistry and Structural Chemistry of LiNiO2(R3m) for 4 Volt Secondary Lithium Cells. Electrochem. Soc. 1993, 147, 1862.
44. Intro to Electrochemical Impedance Spectroscopy (EIS) of Batteries. https: //www. youtube. com/watch?v=xaimI9w-egQ
45. Whittingham, M.S. Lithium batteries and cathode materials. Chem. Rev. 2004, 104, 4271–4302.

Functional Nanomaterials and their 2D and 3D Fabrications for Drug Delivery Applications

Wei-Fan Lee

Grade 10, Taipei Wego Senior High School, Taiwan

Abstract. This review paper would be discussing both the manufacture of the functional nanomaterials and their practical applications for drug delivery The study of nanomaterials has been a coruscating field of research that has a huge impact on science, engineering, and medicine. In terms of applied science, it succeeded the limits of conventional materials, hence making the opt of the materials more multifaceted. Speaking of drug delivery, new research on nanomaterials marked an outset of a new study of medicine and might have changed our perception of drugs. Nanomaterials are recognized to be a emerging material with good prospect for drug delivery regarding their particular characteristics. Because of these benefits, further research on this topic is conducted as of now.

1. Introduction

1.1 Importance of Gold NPs

In the field of drug delivery, NPs are widely used because of their unique properties. Often, NPs are functionalized to meet the requirement for different usages. Among the functionalized NPs, gold particles is one of the best choices among the functionalized NPs because it serves as an excellent agent of carrying anti-cancer during the cancer therapy, and many other significant advantages. It also serves as a dose enhancer during radiotherapy. As Figure 1 shows, the functions of Au NPs can be easily altered by attaching the NP to different substances. This shows that Au NPs are advantageous because of their wide usage.

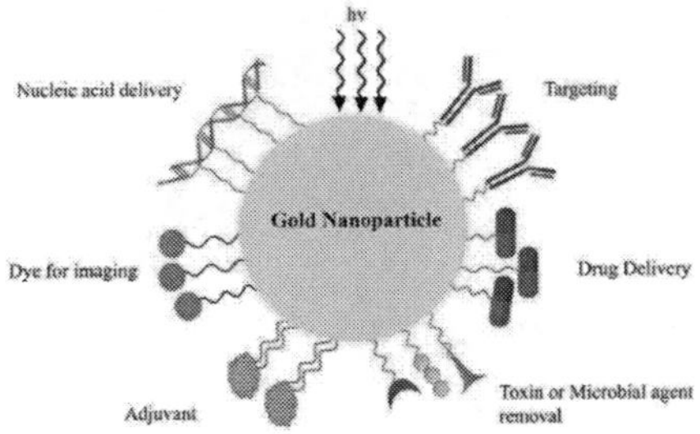

Figure 1 Various functions of Au NPs by altering the materials attached to the NP [1]

As of now, thanks to the advancements in technology, producers now has the ability to fabricate different size, composition, morphology, or other properties by applying precise technology [2]. The chart below demonstrates the variation of Au NPs in terms of properties [2].

Table 1 Variety of properties of Au NPs

Diameter (nm)	Nanoparticles /mL	Peak SPR Wavelength (nm)	Molar Ext ($M^{-1} \cdot cm^{-1}$)	Product Numbers (Surfactant Stabilized)
5	5.47×10^{13}	515-520	1.10×10^{7}	741949
10	5.98×10^{12}	515-520	1.01×10^{8}	741957
15	1.64×10^{12}	520	3.67×10^{8}	777137
20	6.54×10^{11}	524	9.21×10^{8}	741965
30	1.79×10^{11}	526	3.36×10^{9}	741973
40	7.15×10^{10}	530	8.42×10^{9}	741981
50	3.51×10^{10}	535	1.72×10^{10}	742007
60	1.96×10^{10}	540	3.07 x	742015

 https://doi.org/10.1051/matecconf/202135301023

			10^{10}	
80	7.82×10^{9}	553	7.70×10^{10}	742023
100	3.84×10^{9}	572	1.57×10^{11}	742031
150	3.60×10^{9}	Not Measured	N/A	742058
200	1.9×10^{9}	Not Measured	N/A	742066
250	7.1×10^{8}	Not Measured	N/A	742074
300	4.5×10^{8}	Not Measured	N/A	742082
400	1.9×10^{8}	Not Measured	N/A	742090

1.2 Biodegradable nanoparticles

New research in nanotechnology allowed NPs to be applied to medicine. They became an important aspect of therapy since it is potentially beneficial for creating personalized medicine [2]. Among the NPs, biodegradable NPs are vital for the field of drug delivery, since they can be directly consumed without toxicity to the body. Biodegradable NPs like polylactide (PLA) allowed the medicine to be enclosed in the matrix, and because of its small size relative to the cells, NPs are able to travel through the membranes of the cells [44].

1.3 2D nanomaterials in Drug delivery

Graphene, a new 2D nanomaterial discovered in 2005 [3], is one of the nanomaterials with good prospects. Since the carbon atoms in graphene are arranged in numerous hexagons (honeycomb shape) through hybridization bonds, it is able to perform numerous properties. Currently, 2D nanomaterial would be a scintillating new era for the field of drug delivery. Both its physical properties (e.g. size, width) and chemical properties would have a notable impact on the field of drug delivery [43].

2. Gold Nanoparticle

2.1 Properties of Gold Nanoparticles

Gold nanoparticle (Au NPs) has been widely utilized in the field of biotechnology because of its biocompatibility and its several surface functionalities [4]. (Figure 2) They are also beneficial because they can serve as effective radiosensitizers. This property can be extremely vital in the field of drug delivery [5]. Spherical Au NPs have beneficial properties as large surface-to-volume ratio, low toxicity [4].

2.2 Gold nanoparticles-Based Drug Delivery

Au NPs based drug delivery has been one of the scintillating fields of drug delivery regarding its superior performance. Albeit there is no pharmaceutical approval of Au NPs based nanodrugs, much research in Au NPs based drug delivery has been conducted [7]. Au NPs are recognized as effective nano carriers for various kinds of drugs. (Figure 3)
Fucoidan, for reference, is one of immune-therapeutic function polymers that is mostly utilized in pharmaceutical applications. Au NPs that is synthesized with fucoidan perform even better biocompatibility and nontoxicity than those synthesized with naturally occurring polymers (e.g. chitosan, porphyran).

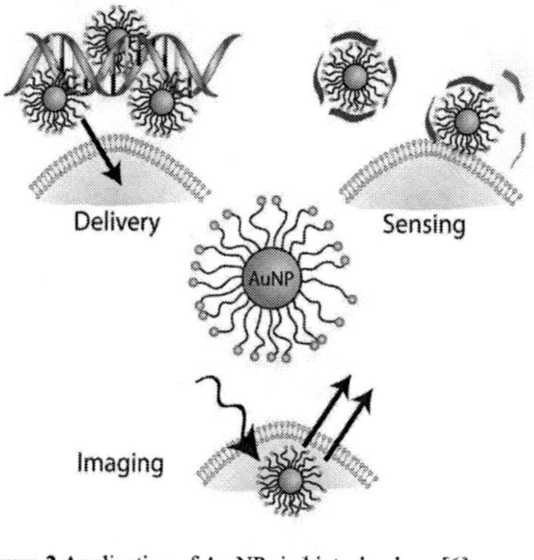

Figure 2 Application of Au NPs in biotechnology [6]

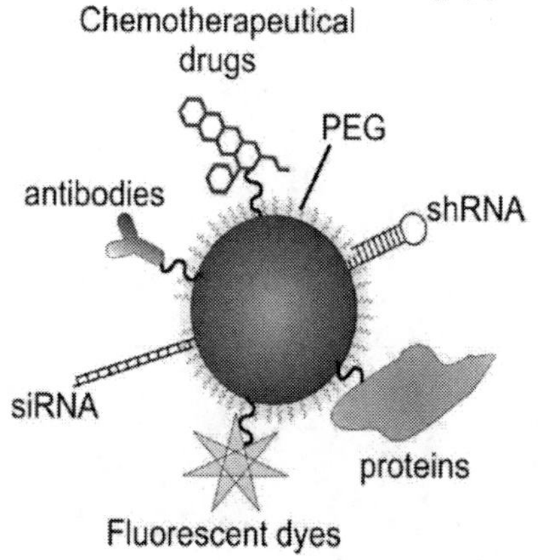

Figure 3 Au NPs functionalization [10]

3. Two-dimensional nanomaterials for drug delivery

The process of drug delivery requires particles, which usually comprise lipids, polymers, or other materials, to

https://doi.org/10.1051/matecconf/202135301023

combine with the therapeutics to substitute the conventional drugs. The free drug system can confront difficulties such as low solubility, loss of drug activity in particular situations. Nano drug delivery focuses on establishing bioactive and biocompatible molecules. Thus, drug delivery systems are expected to evenly release drugs at a designated time [4].

3.1 Graphene

Graphene is a 2D, flat monolayer carbon atoms in the shape of a honeycomb. (Figure 4) It is the basic unit of other graphitic structure (e.g., fullerene (0D), carbon nanotube (1D), and graphite (3D)). Graphene is applied in many research fields due to its advantageous properties (e.g. electric, thermal, optical, magnetic). In the study of biological applications, graphene is also beneficial because of its simple biofunctionalization [4] [11].

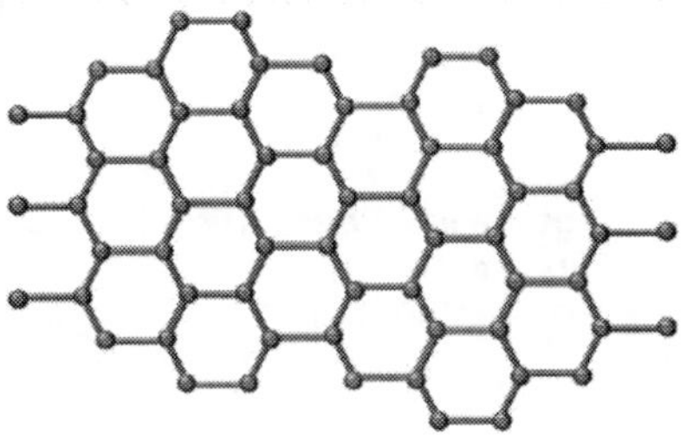

Figure 4 Structure of graphene. Reproduced from [11]

In terms of drug delivery, graphene nanomaterials (GNMs) can be highly beneficial because of their properties (e.g. surface area, layer numbers, lateral dimensions) [12]. Moreover, graphene responds to many environmental factors, including magnetic and electric field, and the existence of enzyme.

3.2 Graphene oxide (GO)

After the oxidation of graphene, numbers of functional group (e.g. epoxide, hydroxyl, carboxylic acid), can be observed on the surface of GO [13]. After the oxidation process, its biocompatibility and combination with biomolecules are enhanced, making it a potential material to be applied to drug delivery [14]. In general, Hummer's method (Figure 8) is usually recognized as the best preparation method of GO since the hydroxyl and epoxide groups can perform a strong π-π bond for drug loading [15].
After synthesis, numerous functional groups (e.g. epoxide, hydroxyl) will be seen on the surface, which can then be used for various purposes [16]. The list below organizes the recent research in the functionalization of GO.

Table 2 Recent research in the functionalization of GO.

Researcher & reference	Research content	Result and prospects
Narayanaswamy et al. [17]	GO-Fe_3O_4	This had led to the depletion of the surface projections. Also, the chromatin condenses and the apoptotic body forms.
Zamani et al. [18]	TiO_2@ZnO-GO	Provided a new method for establishing a hydrophilic drug delivery system

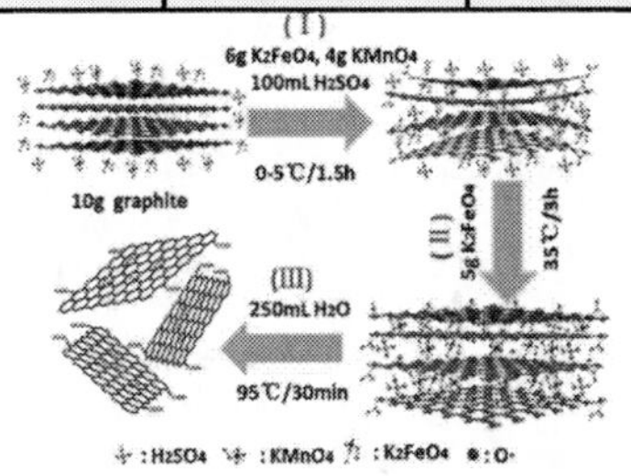

Figure 5 Hummer's method for fabricating GO

3.3 Black Phosphorus

Black phosphorus (BP), another emerging nanomaterial, has the advantage of switchable band gaps, extinct biodegradability, which may be a potential material used for loading biomolecules, drugs, or antibodies [4,21].
BP nanosheets are usually fabricated by exfoliating from BP crystals, which possesses strong interaction between phosphorus and comparatively weak van der Walls force [22]. Its hexagonal honeycomb-like arrangement of the atoms (Figure 9) gives it a semiconductor nature, superior light absorption capability [23] [24].

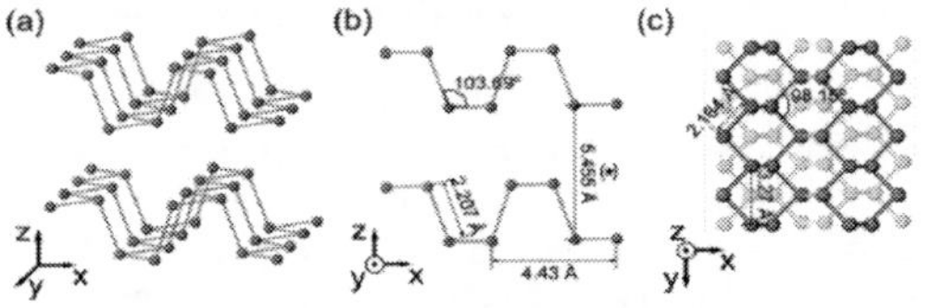

Figure 6 Physical arrangement of BP

BP is rather significant among the nanomaterials used for drug delivery systems (DDSs) due to its high SA/V, photosensitivity, superior biodegradability, and biocompatibility. As shown in Figure 10, there are several categories of BP platforms used for DDSs, including BPN, BPQDs, BPHs [25]. BP has been proving itself to be an

https://doi.org/10.1051/matecconf/202135301023

effective carrier of drugs, RNA interference (siRNA), inorganic substances (e.g., UCNP, Pt) [25].

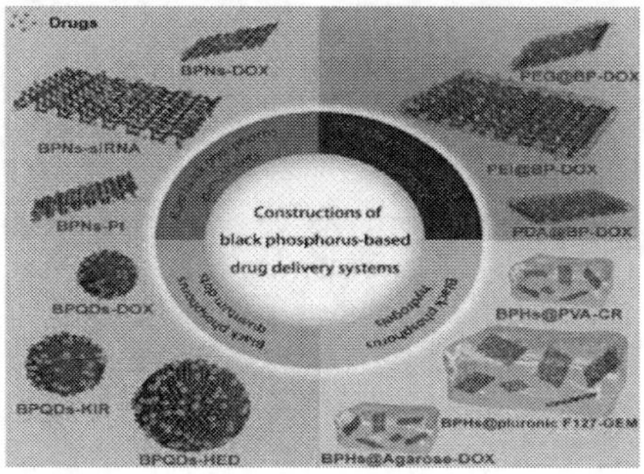

Figure 7 Categories of BP platforms

3.4 Transitional Metal Dichalcogenides (TMDs)

TMDs are a unique kind of compound that is frequently utilized in the field of nanotechnology. Its formula can be presented as MX_2, where M symbolizes the transitional metals, while X represents the chalcogen. The structure of the TMDs usually varies because of different coordination spheres of the metal atoms, which allows diverse loading capability [26]. Conventional TMDs (e.g. MoS_2 , WS_2 , $MoSe_2$, WSe_2), performs the unique structure of planar TMDs crystals.

Thin layer TMDs possess several advantageous attributes which make them potential candidates for the DDSs [27]. In some cases, TMDs perform even more advantageous properties than other 2D nanomaterials. For example, 2D MoS_2 had been proven to have higher loading capacity than GO because of its surface absorption effect [28].

3.5 Transitional Metal Oxides (TMOs)

TMOs are the compound that comprises transitional metal and oxygen. With their 2p metal being filled, TMOs owns a nature of having valence band. The conduction bands can be an important aspect of determining its electric and magnetic properties, which is vital for its application in photochromic and thermochromic related fields [29]. Among the TMOs, several materials as MnO_2 , TiO_2 , ZnO stands out especially since they perform unique ferromagnetic and redox properties, which may be advantageous for constructing photo-sensitizers for DDSs [29].

The redox characteristics of MnO_2 2D nanosheets gives it a signal-responsive nature. With that being said, it is one of the promising candidates to be utilized in cancer therapy [29]. The redox/pH controllable PTX DDSs developed by Choi and colleagues had been proven to reduce many side effects, which can be important for the field of cancer therapy [4].

TiO_2 NPs are also a kind of rising materials because of its semiconductive nature with a wide bandgap [4]. With regard to its 2D characteristic, TiO_2 also has a superior loading capacity. Research conducted by Kushnirov and Sosnik [31] had successfully realized the fabrication of TiO_2/polymer amphiphilic NPs with controllable size and outstanding performances in DDSs.

3.6 Transition metal carbides (TMCs)

Transitional metals carbides (TMCs) are compounds composed of early transitional metals (e.g. Ti, Mo, V, Nb), and C or N [29]. MXenes are potential candidates in many fields including diagnostic imaging, biosafety, issues, making MXenes one of the pioneering nanomaterials among all [33]. As other 2D nanomaterials, MXenes can be functionalized in order to apply it for cancer therapy or other field of application.

Just as TMOs, TMCs also have a large loading capacity of various biomolecules. Research conducted by Han and colleagues had successfully demonstrated that TMCs can effectively load various kinds of organic dyes. To better demonstrate the property of TMCs, Chen and coworkers evaluated its loading capacity, pH/ NIR-triggered release of drugs, PA imaging ability [34]. Results demonstrated the biosafety of MXenes by proving that it can be naturally emitted after a particular time.

4. Types of Biodegradable Nanoparticles and their applications in nanomedicines

4.1 Biodegradable Nanoparticles in targeted Delivery System

Biodegradable NPs used in targeted DDSs can be categorized in regard to their structures or their arrangements just as the nonbiodegradable substances. The wanted substance can either be encapsulated or attached onto the surface of the biodegradable NPs [34]. In general biodegradable NPs possess the conventional advantages NPs like slow and controllable release of drugs. Just as other nanomaterials, biodegradable NPs can be further functionalized to meet different demands [34]. During the process of synthesis, polymers should have natures of biocompatibility and biosafety. Biodegradable NPs, on the other hand, are usually colloidal particles that is available of combining with functional groups. Based on the polymers, polymer-based nanomaterials can be categorized by examining if it is synthesized or natural.

4.2 Antioxidant Delivery

Oxidation form of substances has been associated with the hazardous effects on cells[36]. If the DDSs could tackle with the issue by delivering agents counters oxidative stress, cell wound given rise to by reactive oxygen species (ROS) can be relieved to a certain extent [37].

To realize the idea of antioxidant delivery, Kang and coworker [38] established a new NP, namely PVAX, a substance that comprises copolyoxalate with vanillyl alcohol (VA). When PVAX confronts relatively high concentration of H_2O_2, PVAX would degrade and then release VA, which can serve as an anti-inflammatory and anti-apoptotic substance.

4.3 Oral Drug Delivery

Comparative to injection, orally consumed drug can largely decrease the probability of getting infection and physical discomfort, especially to the diabetic patients [35]. 1.6% Zn insulin in PLGA, for reference, was developed a decade ago to realize the idea of oral drug delivery. Although the practicality of Zn insulin was relatively low, it still paved the way for the future research. Research in oral drug delivery is still being conducted to this day.

5. Conclusion

The article aforementioned had arranged the organized the recent research in functional nanomaterials and their 2D and 3D fabrications for DDSs application. From these topics, one can analyze the vital attributes for nanomaterials to be applied in biotechnology include distinct biocompatibility, biodegradability, non-cytotoxicity.

As of now, nanomaterials have been paving a path for research in many fields. For DDSs, nanomaterials are rising materials to be researched furthermore because of their extinct attributes. The current research progress would not be an terminal to the development of nanomaterials; rather, it marked the outset of another era of DDSs.

References

1. Using gold nanoparticles in diagnosis and treatment of melanoma cancer - Scientific Figure on ResearchGate. Available from:
   https://www.researchgate.net/figure/Different-applications-of-gold-nanoparticles-in-diagnosis-and-therapy-Nanoparticles-are_fig3_322739738 [accessed 5 Nov, 2021]
2. Gold Nanoparticles: Properties and Applications, www.sigmaaldrich.com/TW/en/technical-documents/technical-article/materials-science-and-engineering/biosensors-and-imaging/gold-nanoparticles.
3. Gold Nanoparticles: Properties and Applications, www.sigmaaldrich.com/TW/en/technical-documents/technical-article/materials-science-and-engineering/biosensors-and-imaging/gold-nanoparticles.
4. Mei, Xuan, et al. "Recent advancements in two-dimensional nanomaterials for drug delivery." Wiley Interdisciplinary Reviews: Nanomedicine and Nanobiotechnology 12.2 (2020): e1596.
5. Yeh, Yi-Cheun, et al. "Gold Nanoparticles: Preparation, Properties, and Applications in Bionanotechnology." Nanoscale, U.S. National Library of Medicine, 21 Mar. 2012, www.ncbi.nlm.nih.gov/pmc/articles/PMC4101904/.
6. Siddique, Sarkar, and James C. Chow. "Gold Nanoparticles for Drug Delivery and Cancer Therapy." Applied Sciences, vol. 10, no. 11, 2020, p. 3824., doi:10.3390/app10113824.
7. Farooq, Muhammad U., et al. "Gold Nanoparticles-Enabled Efficient Dual Delivery of Anticancer Therapeutics to Hela Cells." Nature News, Nature Publishing Group, 13 Feb. 2018, www.nature.com/articles/s41598-018-21331-y.
8. Jeong, Eun Hye, et al. "Gold Nanoparticle (Aunp)-Based Drug Delivery and Molecular Imaging for Biomedical Applications." Archives of Pharmacal Research, vol. 37, no. 1, 2013, pp. 53–59., https://doi.org/10.1007/s12272-013-0273-5.
9. Tumor Microenvironment Modulation via Gold Nanoparticles Targeting Malicious Exosomes: Implications for Cancer Diagnostics and Therapy - Scientific Figure on ResearchGate. Available from https://www.researchgate.net/figure/Gold-nanoparticles-AuNPs-functionalization-for-theranostics-After-functionalization_fig1_312481625 [accessed 30 Oct, 2021]
10. Bai, Renu Geetha, and Ghaleb A. Husseini. "Graphene-Based Drug Delivery Systems." Biomimetic Nanoengineered Materials for Advanced Drug Delivery, 2019, pp. 149–168., https://doi.org/10.1016/b978-0-12-814944-7.00011-4.
11. Graphene: Fabrication Methods, Properties, and ApplicatioZns in Modern Industries - Scientific Figure on ResearchGate. Available from: https://www.researchgate.net/figure/Basic-structure-of-graphene-6_fig1_340942598 [accessed 24 Oct, 2021]
12. Liu, Jingquan, et al. "Graphene and Graphene Oxide as New Nanocarriers for Drug Delivery Applications." Acta Biomaterialia, vol. 9, no. 12, 2013, pp. 9243–9257., https://doi.org/10.1016/j.actbio.2013.08.016.
13. Cai, Ren, et al. "Fabrication of Ultrathin Zn(Oh)2 Nanosheets as Drug Carriers." Nano Research, vol. 9, no. 8, 2016, pp. 2520–2530., https://doi.org/10.1007/s12274-016-1138-2.
14. Sharafeldin, Mohamed, et al. "FE3O4 Nanoparticles on Graphene Oxide Sheets for Isolation and Ultrasensitive Amperometric Detection of Cancer Biomarker Proteins." Biosensors and Bioelectronics, vol. 91, 2017, pp. 359–366., https://doi.org/10.1016/j.bios.2016.12.052.
15. Yang, Lingyan, et al. "Photothermal Therapeutic Response of Cancer Cells to Aptamer–Gold Nanoparticle-Hybridized Graphene Oxide under Nir Illumination." ACS Applied Materials & Interfaces, vol. 7, no. 9, 2015, pp. 5097–5106., https://doi.org/10.1021/am508117e.
16. Narayanaswamy, Venkatesha, et al. "Go-FE3O4 Nanoparticle Composite for Selective Targeting of Cancer Cells." Nano Biomedicine and Engineering, vol. 9, no. 1, 2017, https://doi.org/10.5101/nbe.v9i1.p96-102.
17. Zamani, Mostafa, et al. "Mesoporous Titanium Dioxide@ Zinc Oxide–Graphene Oxide Nanocarriers for Colon-Specific Drug Delivery." Journal of Materials Science, vol. 53, no. 3, 2017, pp. 1634–1645., https://doi.org/10.1007/s10853-017-1673-6.

18. Wang, Hui, et al. "Ultrathin Black Phosphorus Nanosheets for Efficient Singlet Oxygen Generation." Journal of the American Chemical Society, vol. 137, no. 35, 2015, pp. 11376–11382., https://doi.org/10.1021/jacs.5b06025.
19. Yu, Huitao, et al. "High-Efficient Synthesis of Graphene Oxide Based on Improved Hummers Method." Scientific Reports, vol. 6, no. 1, 2016, https://doi.org/10.1038/srep36143.
20. Wang, Huaiyu, and Xue-Feng Yu. "Few-Layered Black Phosphorus: From Fabrication and Customization to Biomedical Applications." Small, vol. 14, no. 6, 2017, p. 1702830., https://doi.org/10.1002/smll.201702830.
21. Hu, Guohua, et al. "Black Phosphorus Ink Formulation for Inkjet Printing of Optoelectronics and Photonics." Nature Communications, vol. 8, no. 1, 2017, https://doi.org/10.1038/s41467-017-00358-1.
22. Isolation and characterization of few-layer black phosphorus - Scientific Figure on ResearchGate. Available from: https://www.researchgate.net/figure/Black-phosphorus-structure-Schematic-diagram-of-the-crystalline-structure-of-black_fig8_260483012 [accessed 31 Oct, 2021]
23. Liu, Wenxin, et al. "Current Advances in Black Phosphorus - Based Drug Delivery Systems for Cancer Therapy." Advanced Science, vol. 8, no. 5, 2021, p. 2003033., https://doi.org/10.1002/advs.202003033.
24. Li, Likai, et al. "Quantum Hall Effect in Black Phosphorus Two-Dimensional Electron System." Nature Nanotechnology, vol. 11, no. 7, 2016, pp. 593–597., https://doi.org/10.1038/nnano.2016.42.
25. Mak, Kin Fai, and Jie Shan. "Photonics and Optoelectronics of 2D Semiconductor Transition Metal Dichalcogenides." Nature Photonics, vol. 10, no. 4, 2016, pp. 216–226., https://doi.org/10.1038/nphoton.2015.282.
26. Anju, S., and PV. Mohanan. "Biomedical Applications of Transition Metal Dichalcogenides (Tmdcs)." Synthetic Metals, vol. 271, 2021, p. 116610., https://doi.org/10.1016/j.synthmet.2020.116610.
27. Tian, Bo, et al. "Photothermally Enhanced Photodynamic Therapy Delivered by Nano-Graphene Oxide." ACS Nano, vol. 5, no. 9, 2011, pp. 7000–7009., https://doi.org/10.1021/nn201560b.
28. Song, Fang, et al. "Transition Metal Oxides as Electrocatalysts for the Oxygen Evolution Reaction in Alkaline Solutions: An Application-Inspired Renaissance." Journal of the American Chemical Society, vol. 140, no. 25, 2018, pp. 7748–7759., https://doi.org/10.1021/jacs.8b04546.
29. Choi, G., Piao, H., Alothman, Z. A., Vinu, A., Yun, C.-O., & Choy, J.-H. (2016). Anionic clay as the drug delivery vehicle: Tumor targeting function of layered double hydroxide-methotrexate nanohybrid in C33A orthotopic cervical cancer model. International Journal of Nanomedicine, 11, 337–348.
30. Kulkarni, M., Mazare, A., Gongadze, E., Perutkova, Š., Kralj-Iglicˇ, V., Milošev, I., ... Mozeticˇ, M. (2015). Titanium nanostructures for biomedical applications. Nanotechnology, 26, 062002.
31. Kushnirov Melnitzer, Vladislav, and Alejandro Sosnik. "Hybrid Titanium Oxide/Polymer Amphiphilic Nanomaterials with Controlled Size for Drug Encapsulation and Delivery." Advanced Functional Materials, vol. 30, no. 18, 2018, p. 1806146., https://doi.org/10.1002/adfm.201806146.
32. Soleymaniha, Mohammadreza, et al. "Promoting Role of Mxene Nanosheets in Biomedical Sciences: Therapeutic and Biosensing Innovations." Advanced Healthcare Materials, vol. 8, no. 1, 2018, p. 1801137., https://doi.org/10.1002/adhm.201801137.
33. Lin, Han, et al. "Two-Dimensional Ultrathin Mxene Ceramic Nanosheets for Photothermal Conversion." Nano Letters, vol. 17, no. 1, 2016, pp. 384–391., https://doi.org/10.1021/acs.nanolett.6b04339.
34. Su, Shi, and Peter M. Kang. "Systemic Review of Biodegradable Nanomaterials in Nanomedicine." Nanomaterials, vol. 10, no. 4, 2020, p. 656., https://doi.org/10.3390/nano10040656.
35. Villemin, Elise, et al. "Polymer Encapsulation of Ruthenium Complexes for Biological and Medicinal Applications." Nature Reviews Chemistry, vol. 3, no. 4, 2019, pp. 261–282., https://doi.org/10.1038/s41570-019-0088-0. 38
36. Manke, Amruta, et al. "Mechanisms of Nanoparticle-Induced Oxidative Stress and Toxicity." BioMed Research International, vol. 2013, 2013, pp. 1–15., https://doi.org/10.1155/2013/942916.
37. Lee, D.; Bae, S.; Hong, D.; Lim, H.; Yoon, J.H.; Hwang, O.; Park, S.; Ke, Q.; Khang, G.; Kang, P.M. H2O2 -Responsive Molecularly Engineered Polymer Nanoparticles as Ischemia/Reperfusion-Targeted Nanotherapeutic Agents. Sci. Rep. 2013, 3, 2233.
38. Bae, Soochan, et al. "Hydrogen Peroxide-Responsive Nanoparticle Reduces Myocardial Ischemia/Reperfusion Injury." Journal of the American Heart Association, vol. 5, no. 11, 2016, https://doi.org/10.1161/jaha.116.003697.
39. Using gold nanoparticles in diagnosis and treatment of melanoma cancer - Scientific Figure on ResearchGate. Available from: https://www.researchgate.net/figure/Different-applications-of-gold-nanoparticles-in-diagnosis-and-therapy-Nanoparticles-are_fig3_322739738 [accessed 5 Nov, 2021]

https://doi.org/10.1051/matecconf/202135301024

Development and Suggestions on Container Twist Lock Handling in Automated Container Terminals

Panpan Zhang [1, 2, *]

1 China Waterborne Transport Research Institute, Beijing, China
[2] Key Laboratory of Logistics Equipment & Control Engineering, Beijing, China

Abstract: In this paper automated container terminals were introduced and characteristics of the automated container terminals were summarized. Besides, a common problem of twist lock handling technology in automated container terminal was proposed. different kinds of automatic twist lock handling technologies and equipment were collected and analyzed. Finally, suggestions on twist lock handling were given. With demands of automated container terminals, the twist lock handling technology will still rapidly develop in future.

1 Introduction

Container transportation is a modern advanced transportation mode. Ship transportation, as a main mode of container transportation, has developed rapidly. In recent years, the United States, Britain, Japan and other countries import and about 70 % to 90 % of cargo exported aboard takes advantage of container transportation. The Container terminal shown in Figure 1 is the hub of land-water inter-modal transportation. container terminal plays an important role in the whole world trade process.

Figure 1. The container terminal

Container handling in terminals is usually consisted of ship handling, horizontal transport and yard handling as shown in Fig. 2. Since the birth of the container terminal, the pace of automation has never stopped. Globally, the automation process of container terminals is accelerating.

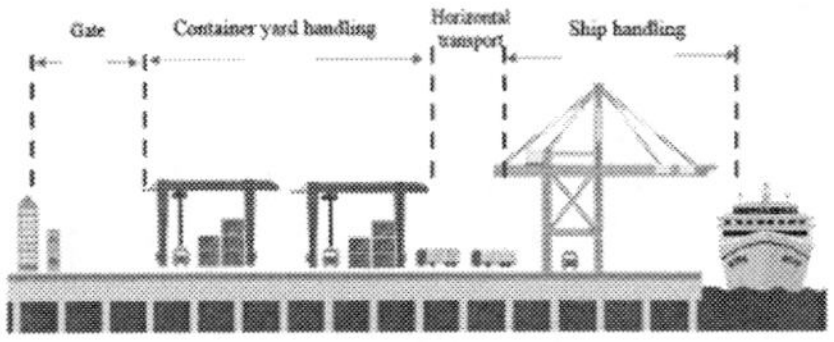

Figure 2. Typical composition of a container terminal

In this paper automated container terminals were introduced and characteristics of the automated container terminals were summarized. Besides, a common problem of twist lock handling in automated container terminal was proposed. Different kinds of automatic twist lock handling technologies and equipment were collected and analyzed. Finally, suggestions on development of twist lock handling technology were given.

2 Characteristics of the automated container terminals

With fast development of the global economy, the maritime container volume around the world shows a growing trend. Container ships are becoming larger and larger. Container handling with stable, efficient, energy-saving, environmental protection and low cost has become the focus of attention of container terminals.

In automated container terminals, ship handling, horizontal transport and yard handling processes can run as the terminal operation system are set. Since the first fully automated container terminal was put into commercial operation in 1993, automated container terminals have developed rapidly. ECT terminal and Euromax terminal at Rotterdam Port, CTA terminal at Hamburg Port, and automated terminal at Qingdao Port as shown in Figure 3 have all become milestones in the development history of fully automated container terminals, witnessing technological changes in container terminal ship handling, horizontal transportation, yard handling and so on. According to incomplete statistics, 10 fully automated container terminals have been built worldwide.

* Corresponding author: zhangpanpan@wti.ac.cn

a) Euromax terminal

b) Qingdao Qianwan terminal

Figure 3. Typical examples of Automated container terminal

1) Automated container terminals play an important role in reducing the labor cost improving the capacity of the port, reducing the energy consumption of the loading and unloading operation. It is an inevitable trend of port development [1]. Characteristics of automated container terminals can be summarized as follows.

2) high reliability:container operations in automated terminal can avoid influence of human factors and can improve the reliability of the operation;

high operation safety: in automated container terminals, all operations were carried out as setting s in terminal system,which results in higher operation safety and is conducive to building safe ports;

3) lower labor costs: automatic operation can reduce the number of personnel assigned to the terminal which is good to reduce labor costs;

4) improvement of the working environment: greatly improve the working environment and reduce labor intensity, reduce occupational diseases, flexible deployment of personnel, shift convenience;

5) low energy consumption and low emission: the extensive use of all-electric drive equipment and the application of energy feedback technology can effectively reduce the energy consumption of port operations and can improve the port environment;

6) larger initial investment: compared with the traditional container terminal, the investment of automatic handling equipment, yard foundation and terminal operation system are higher;

3 Development of Twist lock handling technology

Handling of container twist lock is an important link in container handling process and it is still a bottleneck in automated container terminal. At present, there are many ways of container locking and unlocking in the world, but manual installation and disassembly are adopted as shown in Fig. 4 [2].

Figure 4. Manual handling of twist lock

During ship handling operations, a container vessel with a length of more than 200 meters requires at least 6 to 10 workers in the removal of container locks which seriously restricts the automation of container handling process. Therefore, many research institutions and enterprises in China and abroad have made some progress in areas of twist lock handling. In this part, the research progress of twist lock handling technology at container terminals were comprehensively introduced and technical characteristics were analyzed.

Kalp released a fully autonomous lashing platform as shown in Fig. 5. The platform adopts hydraulic system and the success rate of twist lock handling is more than 95% [3]. The biggest feature of the platform is that twist locks can be removed or installed under different working conditions. When the twist lock is removed, the twist lock is automatically placed in the internal collector. 2,400 twist locks can accommodated in each collector. Besides, the platform is adjustable for all kinds of standard containers.

Figure 5. The autonomous lashing platform

The system can be operated smoothly by integrated hydraulic shock absorption technology as shown in Fig. 6. the design can absorb the force and reduce the impact on cargo and equipment [2]. The energy needed to operate the twist lock handling is generated by the gravity of the container itself. The device is efficient and does not rely on external power (such as fuel and electricity), thus maximizing flexibility for operations.

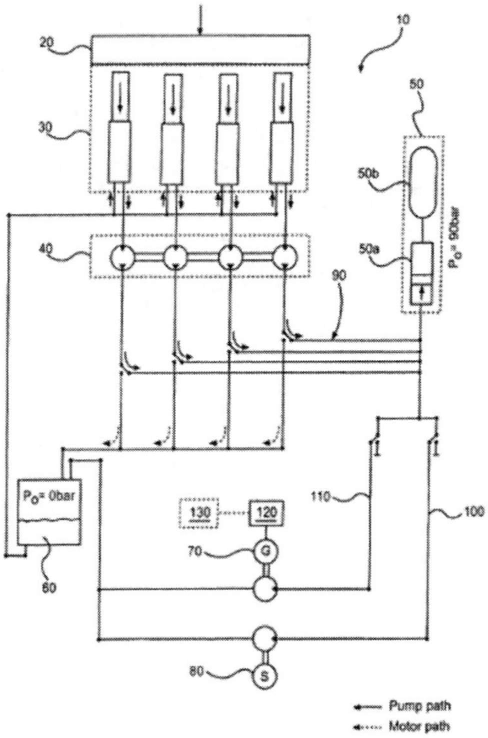

Figure 6. Diagram of the hydraulic system

C. Liang et al. From Nanyang Technological University developed a Grizzly unlocking experimental system based on a manipulator [4]. The system is mainly composed of a manipulator, claws, a camera and a mobile platform. Based on visual recognition, the platform can obtain the position and can recognize the type of the twist lock through the camera. and then the twist lock through the manipulator and three movable claws can be removed. The equipment still needs to be improved to adapt to the on-site situation in the busy horizontal transport conditions of container terminals.

Liu et al. from Qingdao Qianwan Container Terminal proposed an automatic disassembly design for container twist locks [5]. Base on the design, a robotic twist lock handling system were developed as shown in Fig. 7. The system is consisted of six industrial robots, a conveyor line and collecting box. Each robot is equipped with visual recognition and sensors. The optical fiber sensor is used to detect the movement of twist locks and the cylinder. The position of twist locks are determined by CCD and Z direction position sensor, and then the signal is transmitted to the robot controller through image processing to correct the deviation of the unloading position. The twist lock can be removed from the container through the servo torque control. The conveyor line connects the working positions of the six robots to achieve the collecting and distribution of the lock.

Figure 7. Robotic lock removal device and system flow chart of Xinqianwan Container Terminal

Research in the field of automatic twist lock handling was also carried out by China Waterborne Transport Research Institute [6]. An experimental prototype of the automatic handling actuator was designed and manufactured as shown in Fig. 8. The experimental prototype can realize the wireless control of the unlocking device through the controller, and realize the automatic unlocking, manual unlocking and fault feedback and processing. The unlocking time of the automatic unlocking execution device is not more than 5 seconds. Interlocking protection, fault self-diagnosis, warning and treatment measures were integrated in the control system as well as in the remote panel as shown in Fig.9 .

Figure 8. The automatic handling prototype

Figure 9. The remote control panel

The principle of automatic twist lock handling devices mentioned above are similar and a series of industrial tests were carried out. However, the automatic twist lock handling system still can not suitable for all kinds of twist locks. Thus, there are suggestions on twist lock handling in the following part.

1) it is necessary to improve standardization of twist locks. There are regulation on the mechanical properties of twist locks in Standard of ISO1161 and there is no mandatory requirement for the shape of twist locks. This results in a wide variety of twist locks, making the automatic twist lock handling difficult. The standardization of twist locks is the basis for the comprehensive upgrading of automatic twist lock handling technology.

2) with development of image recognition technology and flexible robot technology, the twist lock handling device will be more intelligent. The twist lock handling device currently developed mainly make use of multi-joint industrial robots. In future flexible robots may bring new opportunities to the twist lock handling area.

3) it is necessary to increase the automation degree of twist locks. If the twist lock is automated, handling of twist locks in container vessels will be labor-saving. It is an important aspect of the automated container terminal.

4 Conclusion

With development of global trade, automated container terminals are booming in recent years.In this paper automated container terminals were introduced. Characteristics of automated container terminals were analyzed and summarized. Besides, a common problem of twist lock handling in automated container terminals was proposed. In the automatic twist lock handling area, different kinds of technologies and equipment were collected and analyzed. Finally, suggestions on twist lock handling development were given. With demands of automated container terminals, twist lock handling technology will still develop in future. By solving automation of the twist lock handling process, the potential ability of automated container terminals can be taken into a new stage.

References

1. He Jihong, Application Status and Prospect of Automatic Container Terminal Handling Process System,Port & Waterway Engineering, 2018(6):5-8.
2. Xie Chen, Zhang Panpan, Fei Haibo.Development Trend of Container Twist Locks Technology. China Water Transport, 2015(7):4-7.
3. Rainer Kapelski, fully autonomous lashing platform US Patent No.US9499336,2016
4. Zhang, L., et al.: Twist-lock pose estimation and grasping based on CAD model. In: Proceedings of the 3rd IFTOMM International Symposium on Robotics and Mechatronics, ISRM2013, Singapore, pp. 739–747 (2013)
5. Panpan Zhang, Chen Xie, Haibo Fei. Twist Lock Unlocking Process Research and Unlocking Fixture Design in Container Terminals, Advances in Engineering Research (2015):1122-1126.
6. Liu Yang, Zhang Wei, Zhuang Xinchuan. Design of automatic disassembly device for container twist lock, Port Science & Technology. 2012, 10:27-18.

Stresses and deformations of an eccentric cylindrical cam during hydro-joining

Kelin Wang [1], Wei Zhang [1], Yihua Dou [2], Xiaozeng Wang [3], Yinping Cao [2,*]

[1] Tarim Oilfield Company, PetroChina, Korla,841000, China
[2] Xi'an Shiyou University, Xi'an, 710065, China
[3] Jiaying University, Meizhou, 514015, China

Abstract. The stresses and deformations of an eccentric cylindrical cam during the process of hydro-joining have been calculated in orthogonal curvilinear coordinates according to the mechanical model of the cam and governing equations in terms of appropriate complex potentials with suitable boundary conditions. The radial and shearing stress coefficients determined for an example cam are far less than that of the tangential stress during hydro-joining. The tangential stress coefficient of the outer surface of the cam is greater than that of the internal surface when the polar angle exceeds a particular value. The position of the maximum value of the radial stress coefficient is located on the internal surface of the cam, and the maximum shear stress coefficient is located between the inside and outside surfaces of the cam. The cam deformations on the internal and external surfaces under internal pressure respectively attain maximum values at particular angles. The maximum values of the radial and y-directional deformations are located at the position of the minimum wall thickness. The radial deformations determined for an example cam are far larger that the tangential deformations during hydro-joining. The errors between the theoretical and numerical solutions for the tangential stress and the y-directional deformation are both very small.

Key Words: Eccentric cylindrical cam, stress, deformation, complex potential, orthogonal curvilinear coordinates.

1 Introduction

When the stresses exceed the strength of cam material under the action of the contact pressure, the failure of the cam will occur[1-2]. An eccentric cylindrical cam (ECC) has a cylindrical structure of variable cross-section, and the basic equations governing stresses and deformations are therefore variable coefficient differential equations, which are very difficult to solve[3-4]. To obtain the stresses and deformations of an ECC, many researchers have adopted numerical methods based on the finite element model of the cam[5-8].

In this work, we determine the positions and values of the maximum stress and deformation of an example ECC. We also develop the finite element model of the cam and determine the numerical solutions for the stress and deformation of the cam.

2 Orthogonal curvilinear coordinates

Giving definite constant values to ξ and η, $f_1(x,y)=\xi$ and $f_2(x,y)=\eta$ will represent two curves which will intersect when $f_1(x,y)$ and $f_2(x,y)$ are suitable functions. Different values of ξ and η will yield different curves and a different point of intersection. Thus, each point in the xy plane will be characterized by definite values of ξ and η. ξ and η may be regarded as coordinates of a point. Because given values of ξ and η define the point by means of two intersecting curves, they are called curvilinear coordinates. Polar coordinates are a special case of curvilinear coordinates which have proved very useful for problems of concentric circular boundaries. An ECC has two non-concentric circular boundaries, and such problems typically require curvilinear coordinates, ξ and η,16 defined as

$$z = ia \cdot cth\left(\frac{\zeta}{2}\right) \tag{1}$$

Where a is a real constant, $\zeta=\xi+i\eta$, and $z=x+iy$.

* Corresponding author: caoyinping02@xsyu.edu.cn

3 Stresses of an ECC in orthogonal curvilinear coordinates

3.1 Stress and stain equations in terms of complex functions

Substituting $\xi = \xi_0$ in Equation (1), we obtain $x^2 + (y - acth\xi_0)^2 = a^2 \operatorname{csc}h^2\xi_0$ .The following equations are for the stresses and deformations of an ECC in terms of complex functions in orthogonal curvilinear coordinates.

$$\sigma_\xi + \sigma_\eta = 2\left[\psi'(z) + \overline{\psi'(z)}\right] = 4\operatorname{Re}\psi'(z) \tag{2}$$

$$\sigma_\eta - \sigma_\xi + 2i\tau_{\xi\eta} = 2e^{2i\alpha}\left[\bar{z}\psi''(z) + x''(z)\right] \tag{3}$$

$$2G(u_\xi - iu_\eta) = e^{i\alpha}\left[(3-)\nu/(1+\nu)\overline{\psi}(\bar{z}) - \bar{z}\psi'(z) - x'(z)\right] \tag{4}$$

3.2 Calculation formulas for deformation

Taking the same complex potentials as employed in the calculation of the stress components, we calculate the radial and tangential deformations of an ECC and obtain the expressions for the radial and tangential deformations of an ECC owing to the expansion of the axial tube in the process of hydro-joining.

$$2Gu_\xi = \frac{ap_1}{m}\left[\begin{array}{l} csh(\xi_1-\xi_0)(ch\xi-\cos\eta) - sh^2\xi_0\left(\frac{(1-\nu)}{(1+\nu)}\frac{ch\xi\cos\eta}{ch\xi-\cos\eta}\right) \\ +\frac{sh(\xi_1+\xi_0)}{2sh(\xi_1-\xi_0)}\left(\frac{3-\nu}{1+\nu}\frac{sh2\xi\cos2\eta/2-ch\xi\cos\eta}{ch\xi-\cos\eta} + sh2\xi\cos\eta - sh\xi\right) + \frac{ch(\xi_1+\xi_0)}{2sh(\xi_1-\xi_0)}\left(\frac{3-\nu}{1+\nu}\frac{sh2\xi\cos2\eta+1-2ch\xi\cos\eta}{2(ch\xi-\cos\eta)} + ch2\xi\cos\eta - ch\xi\right)\end{array}\right] \tag{5}$$

$$2Gu_\eta = \frac{ap_1}{m}\left[\begin{array}{l} \frac{ch(\xi_1+\xi_0)}{2sh(\xi_1-\xi_0)}\left(\frac{\nu-3}{1+\nu}\frac{2ch\xi\sin\eta - sh2\xi\sin2\eta}{2(ch\xi-\cos\eta)} + sh2\xi\sin\eta\right) - \frac{1}{2}sh^2\xi_0\left(\frac{2(1-\nu)}{1+\nu}\frac{sh\xi\sin\eta}{ch\xi-\cos\eta}\right) + \\ \frac{sh(\xi_1+\xi_0)}{2sh(\xi_1-\xi_0)}\left(\frac{\nu-3}{1+\nu}\frac{sh\xi\sin\eta - ch2\xi\sin2\eta/2}{ch\xi-\cos\eta} + ch2\xi\sin\eta\right)\end{array}\right] \tag{6}$$

4 Examples

4.1 Stresses of the ECC

As an example, we set the radius of the outer circle of an ECC at 48 mm, the radius of the inner circle at 30 mm, the distance between their centers at 5.4 mm, and the height of the ECC at 18 mm, as shown in Figure 1. The tangential stress coefficients of the internal and external surfaces of the ECC are shown in Figure 2. The shearing stress coefficient at the polar angle π/2 is shown in Figure 3. The radial stress coefficient at the position of the minimum wall thickness is shown in Figure 4. The tangential stress of the ECC at the position of the minimum wall thickness is shown in Figure 5.

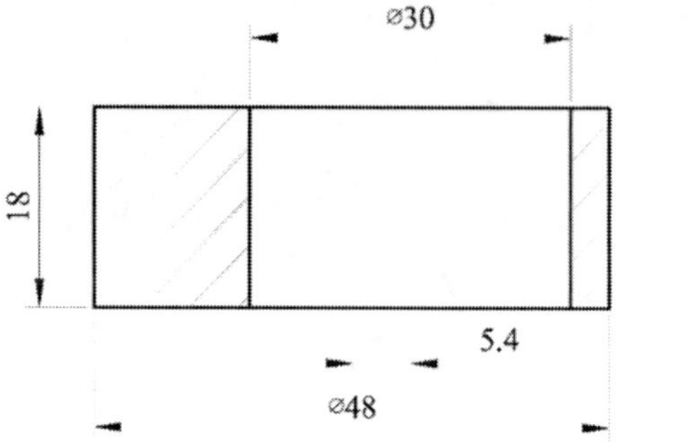

Figure 1. Part drawing of the eccentric cylindrical cam used as an example

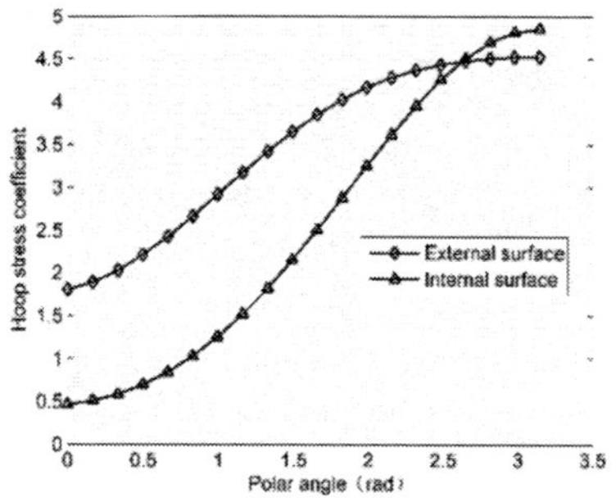

Figure 2. Tangential stress coefficients on the internal and external surfaces of the eccentric cylindrical cam

The tangential stress coefficient is always a positive value, indicating that the tangential stress of the cam is the tension stress owing to the expansion of the axial tube in the process of hydro-joining. The tangential stress coefficient on the external surface of the cam ranges from 0.46 to 4.85 and that on the inner surface ranges from 1.80 to 4.53 as the polar angle varies from 0 to π, which reaches a limiting value at the position of the minimum wall thickness. The tangential stress coefficient on the external surface of the cam is less than that on the inner surface when the polar angle is less than 152°. The radial stress coefficient at the position of the minimum wall thickness varies from 0 to −1 from the external surface to the internal surface of the cam owing to the expansion of the axial tube during hydro-joining. The variation of the tangential stress coefficient is nearly linear. The values of the tangential stress at the position of the minimum wall thickness on the external and internal surfaces of the cam are larger than radial the stress values by factors of 4.85 and 4.53, respectively The shearing stresses at the positions of maximum and minimum wall thicknesses of the cam under uniform inner pressure are equal to zero, and the curve is in the form of a parabola along the minimum wall thickness from the inner surface to the outer surface, which reaches a limiting value at the middle plane of the cam. The shearing stress value is only 0.0776, which is 1.5% of the tangential stress value. The tangential stresses are respectively 242.6 and 226.3 MPa at the internal and external surfaces at the position of the minimum wall thickness of the ECC.

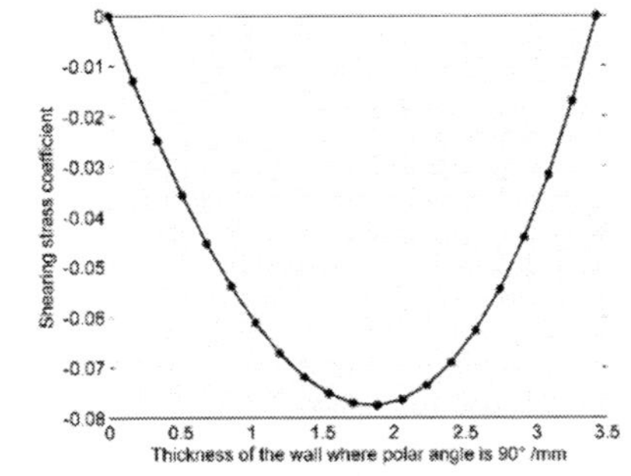

Figure 3. Shearing stress coefficient of the eccentric cylindrical cam at the polar angle $\pi/2$

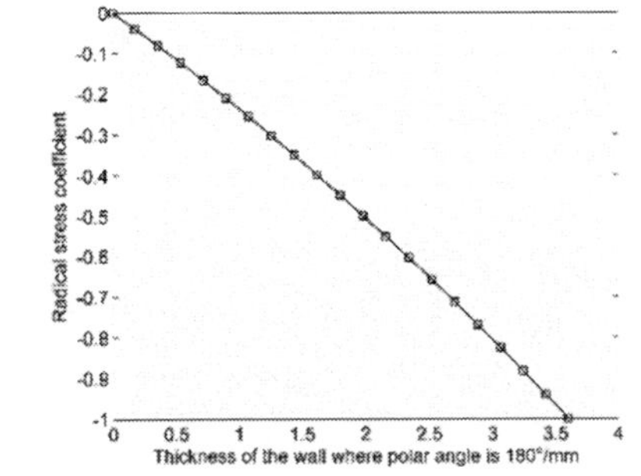

Figure 4. Radial stress coefficient of the ECC at the position of the minimum wall thickness

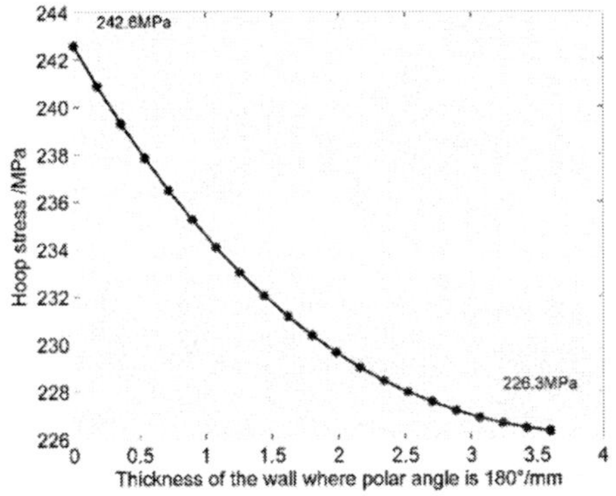

Figure 5. Tangential stress of the eccentric cylindrical cam at the position of the minimum wall thickness

5 Numerical solutions of stresses of the ECC

5.1 Numerical analysis model

The numerical analysis model of the stresses and deformations of the cam employed in the present study adopts the two-dimensional (2-D) PLANE82 8-node element, which is a higher order version of the 2-D, four-node element. The radial and tangential stresses of the ECC under the action of a 20 MPa hydrostatic pressure are shown in Figure 6.

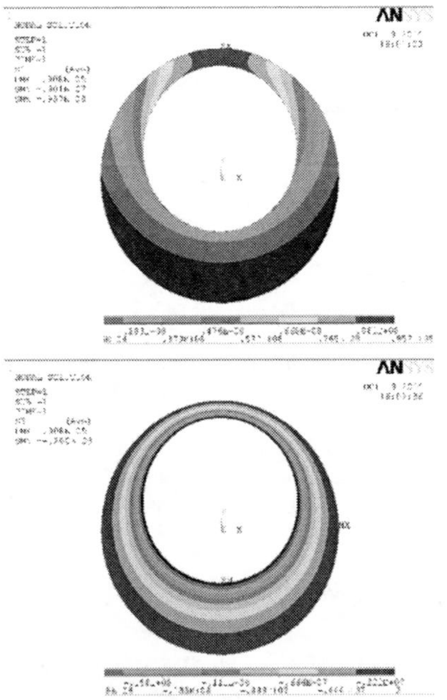

(a) Tangential stress (b) Radial stress

Figure 6. Stresses of the eccentric cylindrical cam under a 20 MPa hydrostatic pressure applied to the internal surface

5.2 Comparisons of the theoretical and numerical solutions of the stresses on the ECC

Theoretical and numerical solutions of the tangential stresses on the internal and external surfaces of the ECC are shown in Figure 12 under the action of various internal pressures during hydro-joining of the cam and axial tube. The tangential stress of the cam is observed to linearly increase with increasing hydrostatic pressure. The theoretical and numerical solutions of the tangential stresses are 243 MPa and 239 MPa, respectively, on the external surface at the position of the minimum wall thickness of the cam, and the error is only 1.6%. The theoretical and numerical solutions of the tangential stresses are 226 MPa and 224 MPa, respectively, on the internal surface, and the error is only 0.9%. For an equivalent uniform internal pressure, the tangential stress on the external surface of is larger than that on the internal surface. The very small errors observed between the theoretical and numerical solutions are the result of the different sizes of meshing employed for the FEM, which can be decreased by meshing a smaller grid in the FEM.

6 Conclusions

The tangential stress and deformation of the ECC are a linear function of the uniform loads acting on the internal surface of the cam during hydro-joining. The radial and shearing stress coefficients determined for an example cam were far less than that of the tangential stress during hydro-joining. The tangential stress coefficient on the external surface was found to be greater than that on the inner surface for a polar angle greater than 152°. The maximum value of the tangential deformations on the internal and external surfaces of the cam are 2.49 and 8.27 μm, respectively, under the action of an internal pressure of 50 MPa, which are located at the polar angles 42° and 62°. Deformations in the radial and y direction at the position of the minimum wall thickness are 23.15 and 21.46 μm, respectively. Deformations in the radial and x direction are 16.59 and 12.15 μm, respectively, which are located in the polar angles 114° and 96°, respectively. Under the action of a hydrostatic pressure of 50 MPa, the differences between the theoretical and numerical solutions of the tangential stress is 1.6% on the external surface of cam, and that of the deformation is 2.0% in the y direction. The errors between the theoretical and numerical solutions of the stresses and deformations of the cam were found to be very small. Therefore, we conclude that the derived theoretical formulas can be applied to the failure analysis of an ECC during hydro-joining.

References

1. Liu Q, Ge JG and Liu G., et al. Analysis of stress and strain of hydro jointing for the PM and tube. Material Science and Technology 2007; 15: 614-618.
2. Lin YH, Qi X and Deng KH, et al. Emperical formula given for collapse resistance of worn casing under nonuniform loading. Oil Gas J 2011; 4: 60-64.
3. Zhou ZF, Wang XY and Gu JF. Numerical Simulation of Eccentric Cylinder Quenching Process. J Mech Eng 2011; 47: 62-67.
4. Zhong RB, Zhou J and Liu WH, et al. Theoretical investigation of a terahertz transmission line in bipolar coordinate system. Sci China 2012; 55: 35-42.
5. Zenkour AM. Stresses in rotating heterogeneous viscoelastic composite cylinders with variable thickness. App Math Mech 2011; 32: 507-520.
6. Liu MT, Li G, Li YQ, et al. Contact analysis between cam and rollers of inner-displaced indexing mechanism based on FEM[J]. Journal of Machine Design, 2015, 32(4): 14-18.
7. Wang XZ. Analysis on hydro-joining stress and deformation of eccentric cylindrical cam and tube[J]. Journal of Machine Design, 2019, 36(02): 88-95.
8. Dong MW, Shu X, Dai MH. Research on Modification Method of Convex Profile of Cylindrical Cam Continuous Transmission Mechanism [J]. Hoisting and Conveying Machinery, 2017(03): 26-29.

The Preparation of Boron-doped Activated Alumina by impregnating boron in the period of boehmite

Rong Fu*

Department of Applied Chemistry, Liaoning Petrochemical Vocational Technology College, Jinzhou, Liaoning, China

Abstract. This paper systematically studied the pore structure, crystal phase, surface acidity and the chemical environment of aluminum of boron-doped γ-alumina, by impregnating boron in the period of boehmite. The nitrogen adsorption, XRD, FTIR etc. technique was used to determine the pore texture, crystallography, acid/base properties and aluminum coordination. the XRD spectrum of boria-alumina have no significant change due to the addition of boron. The specific surface area and pore volume diameter of B-Al2O3 increase significantly. The NMR spectra of B-Al2O3 shows that boron changed the chemical environment of aluminum, which caused differences in the physical and chemical properties of the surface.

1 Introduction

Alumina is called activated alumina due to its special surface properties, as well as its thermal stability, also it has surface acidity and adjustable pore structure, so its applications in the fields of catalytic materials, drying materials, adsorption materials are very significant. Especially in the high temperature reaction fields, such as the oil hydrogenation cracking, hydrogenation and dehydrogenation, catalytic purification of automobile exhaust, γ-alumina is widely used as a catalyst support. However, γ-alumina which has the cationic defects in its spinel structure, belongs to the transitional type of metastable phase. At present, researches of the preparation of modified γ-alumina are most focused on sol-gel method[1,2], precipitation method, microemulsion method , combustion synthesis[3] etc, but all these methods have its only disadvantages.

As the quality of crude oil declines, while the existing hydrofining catalyst by changing, the process conditions can be produced the products to meet the standard of euro VI gasoline and diesel oil. But a series of problems emerges, such as reaction severit, high energy consumption and low selectivity. The fundamental ways to solve these problems was to develop new type of efficient hydrogenation catalysts with high intrinsic activity and selectivity. To solve this problem, we must understand deeply of some key scientific problems, The following paper will systematically examine the effect of boron on the physical and chemical properties of activated alumina and provide a reference for further in-depth study on hydrogenation catalysts.

2 Experimental

Aluminum nitrate and ammonia water are selected as the aluminum source and the alkali source, respectively. After neutralization to form the aluminum hydroxide gel, Continue to add ammonia, adjust the pH of the sol-gel to 8. after stirring for 2 h, let stand at room temperature for 24 h, then dry it in the oven for 24 hours at 80 ℃. The sample is synthesized by impregnating H3BO3 with equal volume in the pseudo-boehmite stage. the boron-doped activated alumina oxide is made by aging and drying . It calcined for 3 h in flowing air at 550°C finally. The prepared samples were called boria-alumina with being the atomic B:Al ratio of 0.02.

3 Results and discussion

3.1 Crystal phase and morphology

Figure 1 shows the XRD of pure alumina and boron-doped alumina. It can be seen from the figure that there were obvious characteristic diffraction peaks of γ-Al2O3 at 37.5°, 46.2° and 66.4°.

At this time, there was still no characteristic peak of boron-containing species in the XRD spectrum, indicating that phosphorus it still exists in an amorphous state or the boron compound is highly dispersed in alumina. Doped boron does not affect the structure of activated alumina. The characteristic peak intensity of γ-Al_2O_3 phase is reduced, indicating that its crystal phase structure is incomplete and its crystallinity is poor.The result is that the alumina surface is covered with a layer of amorphous borate, which causes the intensity of the alumina diffraction peak to decrease.

* Corresponding author: 851360528@163.com

https://doi.org/10.1051/matecconf/202135301026

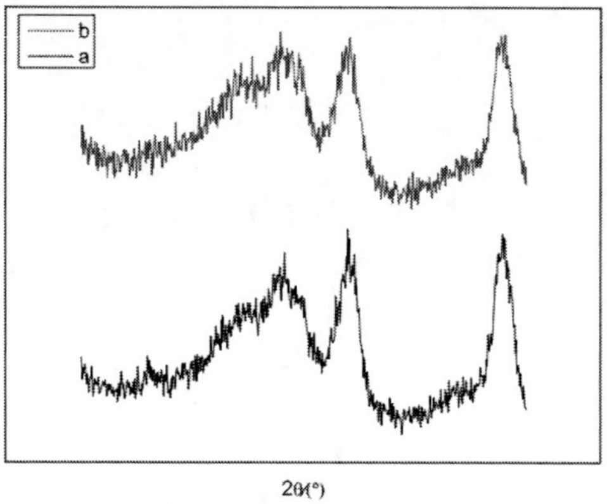

Figure 1. The XRD of pure alumina (a) and boria-alumina (b)

The modified γ-Al2O3 powder is in irregular particle shape, with a relatively uniform particle size distribution, with a particle size of about 20 nm, which is larger than the average particle size calculated by Scherrer's formula. This is because the particles have high surface energy and attract each other, resulting in a certain agglomeration of the powder. Therefore, the result of electron microscopy characterization is slightly larger.

Figure 2. The TEM of boria-alumina.

3.2 Surface area and porosity

Figure 3 shows the adsorption isotherms of the boria-alumina sample and pure alumina. After calcination, the adsorption isotherms are similar in shape, which are all ink bottle-shaped pores, indicating that the pore type of the sample is not affected by the boron content.The characteristic of this type of hole is that the size of the orifice is similar to that of the cavity, which is generally a hole formed by the accumulation of particles. The reason is that the hole of the ink bottle-shaped hole is small, which is easy to be closed due to high temperature sintering. At the same time, the particles grow and accumulate with each other, so the hole type turns into a slit hole.

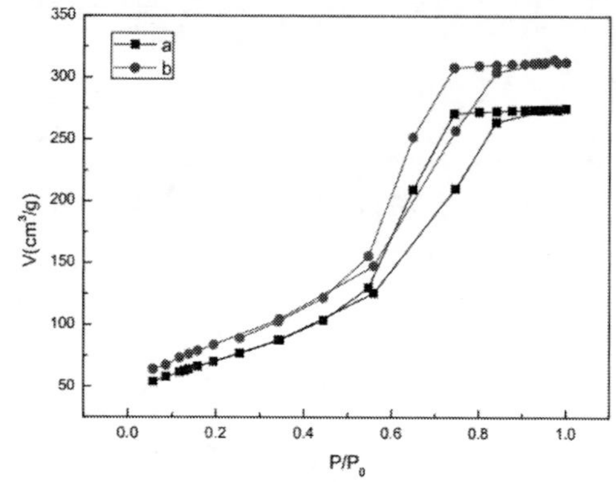

Figure 3. Nitrogen adsorption-desorption isotherms of pure alumina (a) and boria-alumina (b)

Table 1 shows the specific surface area, pore volume and average pore of the pure alumina and the boria-alumina sample. The addition of boric acid did not significantly change the pore size of activated alumina, but compared with pure alumina. Specific surface area increased by 19%, the pore volume increased by 14%, and the average pore width decreased by 4%.

Table.1 The surface area and porosity of pure alumina and boria-alumina

	pure alumina	boria-alumina
surface areas (m^2/g)	256	305
pore volume (cm^3/g)	0.43	0.49
average pore (nm)	6.7	6.4

3.3 Surface acidity

The Table 2 shows the surface acidity of B-Al_2O_3. In boron-doped activated alumina, strong acid and medium strong acid account is more, while weak acid account is less.

 https://doi.org/10.1051/matecconf/202135301026

Table.2 The surface acidity of B-Al_2O_3

	weak acid sites /%	medium strong acid sites /%	strong acid sites /%	Total acidity mmol/g
B-Al_2O_3	38.59	39.75	21.66	0.724

3.4 The chemical environment of aluminum

There are three states of aluminum in the aluminum spectrum[4-6], which are distributed as characteristic peaks near chemical shifts 0, 30, and 60 ppm, which represent six-coordinate aluminum (AlVI), five-coordinate aluminum (AlV), and four-coordinate aluminum (AlIV). It can be seen from Figure 4 that the boron-doped activated alumina has characteristic peaks of six-coordinate aluminum and four-coordinate aluminum around chemical shifts of 8 ppm and 65 ppm.
After fitting the peak to the sample, there is a fitting peak at δ=30 ppm. The characteristic peak near 30 ppm belongs to five-coordinate aluminum, indicating the presence of five-coordinate aluminum ion (AlV) in this sample. This may be due to the dehydration reaction between boric acid and pseudo-boehmite, which distorted the crystal structure of pseudo-boehmite on its surface, and the crystal became incomplete, resulting in more AlVI ions exposed to OH.

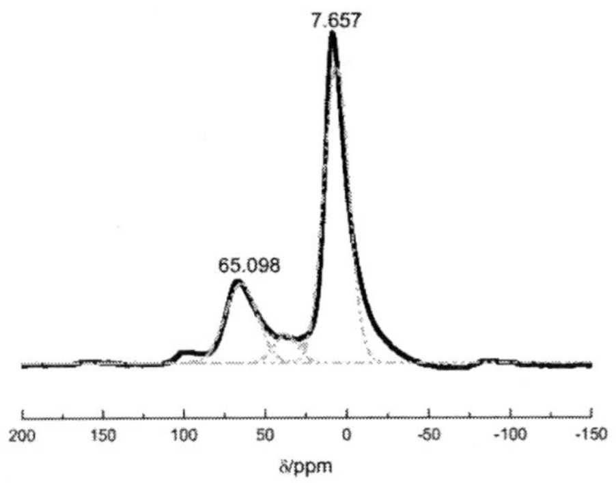

Figure.4 The 27Al CP/MAS NMR spectra of B-Al_2O_3

In the boron-doped activated alumina sample, the ratio of Al^{VI}/Al^{IV} is 3.57/1, which exceeds the theoretical value. It is speculated that boron occupies tetrahedral sites in the form of boron-oxygen tetrahedrons, so that during the transformation of the precursor to γ-Al_2O_3, the migration of Al^{VI} to Al^{IV} is inhibited, and the number of Al^{IV} is reduced.

Table.3 The Effect of Boron Doping on fitting profiles of 27Al NMR

sample	center	width	height($\times 10^8$)	area ($\times 10^9$)	Al^{VI}/Al^{IV}
pure alumina	6.857	15.230	3.815	7.283	2.50
	64.921	19.937	1.163	2.905	
boria-alumina	4.643	17.395	4.016	8.755	3.57
	64.959	15.359	1.273	2.451	
	44.876	18.137	0.451	1.026	

4 Conclusion

Boron-doped Activated Alumina was impregnated in the period of boehmite. XRD results showed that the boron-doped activated alumina will not form a new phase, but reduce the intensity of the diffraction peak. NH3-TPD results showed that the acidity is mainly the medium strong acid and the weak acid and the specific surface area increase. The result of NMR indicates that boron changes the hemical environment of aluminum, then causes the differences in physical and chemical properties of the surface.

Acknowledgments

The author thank technology plan project No. 19B1D27 and 2021 Basic Scientific Research Project of Higher Education Institutions of the Education Department of Liaoning Province (General Project), which have financial supportted this work.

References

1. Shi T, Guo X Z, Yang H. Preparation and Characterization of Nanocrystalline γ-Al2O3 by Sol-Gel Process[J]. Rare Metal Materials and Engineering, 2008, 37(Suppl.2): 379 ~ 382.
2. Tian X S, Ren S X, Wang Q P. Alumina Membrane Modified by ZrO2 and SiO2[J]. China materials technology and equipment, 2007, (4): 46 ~ 49.
3. Patil K C, Aruna S T, Tanu MInani. Combustion synthesis[J]. Current Opinion in Solid State and Materials Science, 1997, 12(2): 158 ~ 165.
4. Huggins B A, Ellis P D. Aluminum-27 nuclear magnetic resonance study of aluminas and their surfaces[J]. J Am Chem Soc, 1992, 114 (6): 2098~2108.
5. Krokidis X, Raybaud P, Gobichon A E, et al. Theoretical study of the dehydration process of

 https://doi.org/10.1051/matecconf/202135301026

boehmite to γ-alumina[J]. J Phys Chem B, 2001, 105 (22): 5121~5130.

6. Damodaran K, Rajamohanan P R, Chakrabarty D, et al. Triple-quantum magic angle spinning 27Al NMR of aluminum hydroxides[J]. Journal of the American Chemical Society, 2002, 124(13): 3200~3201.

AUTHOR INDEX

EDP Sciences – Web of Conferences
17, Avenue du Hoggar
Parc d'Activité de Courtabouf
BP 112
F-91944 Les Ulis Cedex A
France

ISBN 978-1-7138-4183-8

CURRAN ASSOCIATES INC.
proceedings.com

9 781713 841838